Biomimetic Design Method for Innovation and Sustainability

Yael Helfman Cohen · Yoram Reich

Biomimetic Design Method for Innovation and Sustainability

Yael Helfman Cohen
Tel Aviv University
Tel Aviv
Israel

Yoram Reich
Faculty of Engineering
Tel Aviv University
Tel Aviv
Israel

ISBN 978-3-319-81651-7 ISBN 978-3-319-33997-9 (eBook)
DOI 10.1007/978-3-319-33997-9

Softcover reprint of the hardcover 1st edition 2016

Printed on acid-free paper

This Springer imprint is published by Springer Nature
The registered company is Springer International Publishing AG Switzerland

Preface

Biomimicry is captivating. It evokes and attracts the interest of people around the globe, from various disciplines, ages, and professions. It is made of the right ingredients: colorful images, nature, inspiration, innovation, and success stories. In technology, biomimicry provides us with amazing materials and structures and suggests solutions beyond our regular thinking patterns. It is a promising path to address some of the major sustainability challenges of humanity today. In science, biomimicry is a fascinating area for study. It is still in formation and therefore leaves ample space for creation and contribution. It involves many disciplines that should be integrated wisely and consequently challenges the disciplinary nature of science.

I (Yael) was exposed to biomimicry as a child during the eighties, when I got as a present the fascinating book "Bionics-Nature's patents." At that time, the prevailing term was bionics. The book waited on my shelf for almost 25 years until it became the beginning of my new career, as biomimicry researcher and consultant. As an adult, already experienced engineer, I discovered again the biomimicry field, this time under the name biomimicry, during one of my design projects. At that time, biomimicry was brand new in Israel with zero information in Hebrew. I was captivated by the endless inspiration that nature could provide to technology. Next, I became a cofounder of Biomimicry IL, a not-for-profit organization that spread the biomimicry seeds is the industrial, academic, and educational sectors in Israel. Not much later, I associated with Prof. Yoram Reich, and we began our biomimetic journey: Yoram as a supervisor and myself as a Ph.D. candidate. This book is the result of this journey, and it is mainly based on my doctoral thesis but includes also additional insights from our backgrounds in design theories and biomimicry practice.

When we first approached the field few years ago as scientific researchers and practitioners, we realized that below the appealing magic, there was almost no scaffolding to lean on. It was not clear what knowledge bases could support the conjunction of distant disciplines and what language should be used for this purpose.

Motivated by the appeal of biomimicry, and by the lacuna of practical biomimetic design methods, we aimed to develop a new biomimetic design method. We wanted to promote the scientific understanding of the field on one hand, but to provide useful method for practitioners on the other. It was clear that we first needed to develop the missing scaffolding: some solid knowledge bases and multidisciplinary language. Inspired by the theory of inventive problem solving (TRIZ), based on identification of recurring patterns in various disciplines, we intuitively believed that patterns may be the basic words of the missing language. We went out for our patterns journey and became "patterns hunters," looking for design patterns that emerge from a large number of biological design solutions: structure–function patterns and sustainability patterns. From this moment, the patterns were the missing scaffolding and the new language to sustain the development of the new design method: the structural biomimetic design method. We invite you to join this journey and share with you our enthusiasm and insights.

Scientists in the field of biomimicry will find an extensive literature review about the biomimicry discipline including detailed review of current biomimetic design methods and tools, and a mapping of research gaps and challenges. Scientists in the field of design theories will find a unique documentation of design method formation accompanied with a detailed model for "Designing a design method."

Practitioners will find a comprehensive design algorithm and practical tools to lead biomimetic design processes, including detailed case studies. Practitioners with special interest in sustainable design will find a bioinspired sustainability tool, the ideality tool, which can be integrated within biomimetic design processes or stand-alone as a sustainability tool.

The book could be used in an undergraduate or graduate course on biomimicry, design theory, product design, or sustainability to provide in-depth material on the subject.

Acknowledgements

This book presents the results of research made possible by the interdisciplinary environment at the Porter School of Environmental Studies at Tel Aviv University. The authors also thank the following individuals for their contribution:

The engineering students at the mechanical engineering program who participated in the sustainability experiments.
The M.Sc. students Yoav Miraz and Ziv Nahari for the development of the case studies of the method.
The innovation experts Avi Sheinman and Amos Redlich for their help in assessing the results of the innovation study.
Alon Weiss for his assistance with the ideality tool case study.
Dr. Sara Greenberg for her assistance with TRIZ-based analyses and models.

I thank my parents Jacob and Dorit Helfman for providing me the seeds of curiosity; my husband Eytan for the support, confidences, and patience; and my beloved children, Maya, Gili, Aviv, Daniel, and Eyal for being my source of love and motivation.

I thank the Porter School of Environmental Studies, Tel-Aviv University, for supporting this study and providing an interdisciplinary research climate.

Yael Helfman Cohen

I thank my parents Rina and Yoseph Reich, for the strong roots and lifetime support; my children Clil (Arbol Del Amor) and Shaked (Almond), for carrying my fascination of nature in their names and personality; and my partner Nurit, for creating a nourishing, sustainable environment.

Yoram Reich

Contents

Part IV Experimentation

Part V Epilogue

Abbreviations

ACC	Amorphous calcium carbonate
AD	Axiomatic design
ARIZ	(English: algorithm for inventive problem solving)
ASIT	Advanced systematic inventive thinking
BID	Bioinspired design
CTO	Chief technology officer
DANE	Design by Analogy to Nature Engine
DFE	Design for the environment
E2B	Engineering to biology
FBEI	Fermanian Business and Economic Institute
GNU	Recursive acronym for "GNU's Not Unix!"
IEKG	Interdisciplinary engineering knowledge genome
ISO	International Organization for Standardization
MEMS	Microelectro mechanical system
NID	Nature inspired designs
ns	Not significant
PROSA	PRoduct Sustainability Assessment
PSI	Product, Social, Institutional
SAPPhIRE	State-Action-Part-Phenomenon-Input-oRgan-Effect
SBF	Structure behavior function
SEM	Scanning electron microscope
Su-Field	Substance-field
TRIZ	Teoriya Resheniya Izobretatelskikh Zadach (English: theory of the resolution of invention-related tasks)
VAS	Visual analogue scale
VDI	Verein Deutscher Ingenieure (English: Association of German Engineers)
WWH	What–why–how

List of Figures

List of Tables

How to Read the Book

Chapters 1–3 provide an extensive introduction to the biomimicry discipline, including review of design processes, methods, and tools. It is recommended for both scientists and practitioners who are looking for a structured review.

Scientists/researchers who are looking for research directions and ideas may find interest in Chap. 4, a mapping of the research gap and challenges, as well as in the summary of future research in Chap. 13. Researchers in the field of design theories or biomimetic design will benefit from Chap. 5, a presentation of a research model: how to design a design method. They may also benefit from Chap. 6, a review of theories, knowledge bases, and conceptual frameworks that may serve as a foundation for biomimetic design scaffolding. Detailed description of the research methodology, process, and results appears in Chaps. 7–9. Chapter 12 has value for researchers who are interested in assessing design methods by field and class experiments. Innovation researchers may find interest in innovation assessment, criteria, and rubric, presented in Chap. 11, as well as in a review of the innovation aspects of this research, in Chap. 13.

Practitioners who want to gain practical tools should first focus on Chaps. 7–9 in order to understand the knowledge bases of the structural biomimetic design method. The core of the practical knowledge is located in Chap. 10, presenting in details the design algorithm and tools to support the structural biomimetic design process. Some of these tools stand-alone and may be used during general biomimetic design processes, not only with the structural design process. Special attention should be devoted to Chap. 11, which includes four detailed case studies from biology to an application and vice versa. Engineers may find special interest in this design method as it is based on the TRIZ knowledge base and views biological systems as if they were technological systems.

Sustainability researchers and practitioners of sustainable design will find interest mainly in Chap. 9. This chapter elaborates on sustainability patterns in nature and presents the ideality bioinspired tool for sustainable design.

Part I
Introduction

Chapter 1
The Biomimicry Discipline: Boundaries, Definitions, Drivers, Promises and Limits

1.1 The Origins of the Biomimicry Discipline

Since the dawn of history, human beings have observed nature and applied its lessons. The significant modeling advances of recent decades and the ability to examine natural solutions by technological means such as scanning electron microscope (SEM), have promoted the expansion of study and practice of nature. In the middle of the 20th century, the idea that new technologies can benefit from biological knowledge pervaded the scientific community and began to consolidate as a distinct domain of research and application: Biomimicry (Bio = life, Mimicry = imitate/mimesis).

Biomimicry is an intended emulation of nature life solutions for solving contemporary challenges. It is based on viewing 3.8 billion years of evolution as a "design lab" and observing its results. Nature serves as a model, mentor and measure for promoting sustainable innovation designs [1], rather than only a source of materials.

1.2 The Biomimicry Discipline—Boundaries, Terminologies and Research Scope

Biomimicry is part of a general trend of convergence between biology and engineering. This convergence is becoming a source of innovation of the twenty-first century. It offers opportunities for mutual knowledge transfer, potential emergence of a new body of knowledge, as well as transfer of actual substances between the domains.

Focusing on the biomimicry discipline, there are also various definitions including biomimetics, bionics, bio-inspiration and bio-inspired design (BID). In some countries, all these terminologies are considered synonyms, but in others, these terminologies are understood differently. The distinction between these bio-terms is not always clear and may confuse both newcomers and experienced researchers or practitioners.

Y.H. Cohen and Y. Reich, *Biomimetic Design Method for Innovation and Sustainability*, DOI 10.1007/978-3-319-33997-9_1

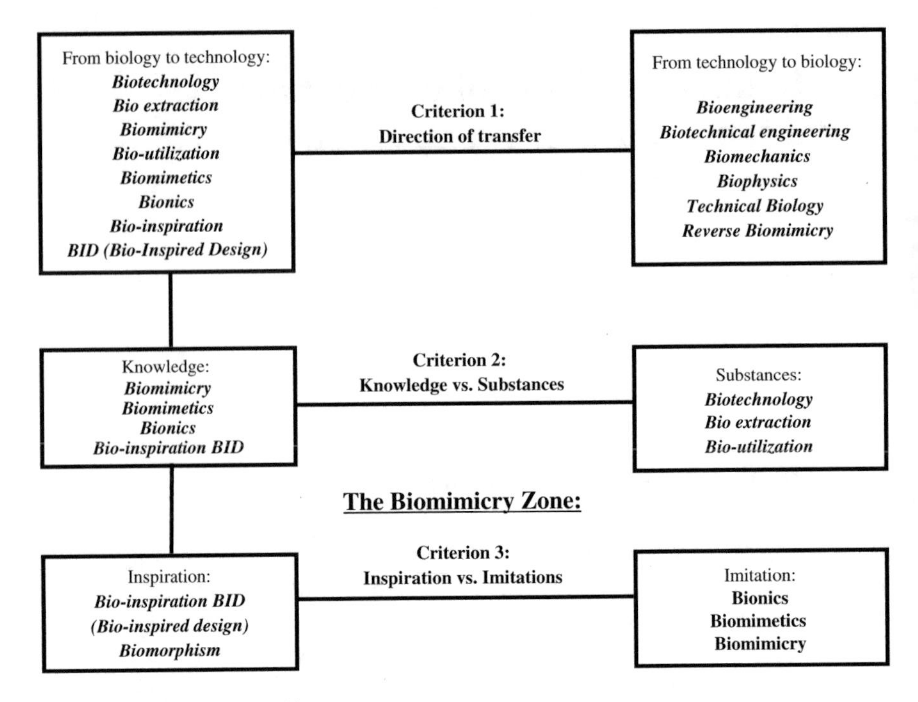

Fig. 1.1 Road-map for the bio-terms

The purpose of this section is to delineate the borders of the biomimicry discipline in relation to closely related disciplines. Based on literature definitions, we suggest three distinctive criteria between the bio-terms. In Fig. 1.1 we present how each criterion narrows the space of bio-terms by excluding several bio-terms, till we reach finally the biomimicry zone.

1.2.1 Criterion 1: Direction of Transfer

We distinguish between transfer from biology to technology or vice versa [2]. We exclude from the biomimicry zone bio-terms that refer to transfer of knowledge from engineering or technology to biology including:

- *Bioengineering*, which is based on the application of engineering knowledge to the fields of medicine or biology; it is also called *biomedical engineering* (Webster's Dictionary). A similar definition is *Biotechnical engineering* that involves the application of engineering principles and tools to solve problems in life sciences [2].
- *Biomechanics*, which is the application of mechanical principles to study and model the structure and function of biological systems [2].

- *Biophysics*, which also refers to the application of physical sciences to solve problems in biological sciences [2].
- *Technical biology* was defined by Nachtigall [3] as the process of using technology for understanding nature. A similar term is *Reverse biomimicry* that is relevant when the biomimetic process itself gains new knowledge about the observed biological system.

1.2.2 Criterion 2: Knowledge Versus Substances

We distinguish between using biological substances (using nature literally) and using biological knowledge (learning from nature). We exclude from the biomimicry zone bio-terms that are related to the usage of biological substances including:

- *Biotechnology*, which is the manipulation of living organisms or their components to produce products (Modified from Webster's Dictionary).
- *Bio-extraction* or *Bio-utilization* that refer literally to extraction of living sources for various uses including for creating products.

1.2.3 Criterion 3: Imitation Versus Inspiration—The Biomimicry Zone

Some of the biological mechanisms are imitated straightforwardly (biomimicry, biomimetic, bionics), while others tend to inspire design concepts (bioinspiration, BID—bio-inspired design) and their relation to biological models may be only loose. On one hand, straightforward mimicking of a biological system can be difficult due to various limitations such as scale, materials and manufacturing capabilities, suggesting that nature should be more a source of inspiration. On the other hand, nature may provide more than just inspiration but full models that could compensate for mathematical limitations, such as in the case of fluid mechanics models [4]. For example, the kingfisher beak provided a model for the optimal shape of the bullet train in Japan [5], instead of developing this model from scratch. Bio-inspiration is more related to transferring of ideas or general design principles while imitation involves transferring of more detailed knowledge including models and exact parameters. Another bio-term that is relevant to this discussion is biomorphism that is mainly related to the art and architectural disciplines. It refers to the processes of designing based on nature patterns or shapes. One example is the Sagrada Família church by Antoni Gaudí that contains many shapes inspired by nature, such as branching "trees" columns. The level of learning may be aesthetic or functional and it is usually more related to bio-inspired processes.

1.2.3.1 Differences Between: Biomimicry, Biomimetics, Bionics

These terms are usually considered synonyms, but as they were not evolved at the same time, they represent different points of development of the biomimicry discipline. The term *bionic* was offered first by Jack Steele, during a US air force symposium, in 1960 [6]. It was defined as the science of systems that include some **functions** copied from nature, or which represent **characteristics** of natural systems. Later (1969), the term *biomimetics* was offered by Otto Schmitt, and defined as the study of the **formation**, **structure**, **or function** of biologically produced substances and materials and biological **mechanisms and processes**, especially for the purpose of synthesizing similar products by artificial mechanisms which mimic natural ones (Modified from Webster's Dictionary). Here we refer also to structures, mechanisms and processes and not only to functions.

And last appeared the term *biomimicry* (1982) [7] but it was widely spread years later (1997) by Jenin Benyus in her book "Biomimicry-innovation inspired by nature" [1]. *Biomimicry* and *biomimetics* are derived from the Greek words bio (life) and mimesis (imitation). The latter was coined by Aristotle who was interested in imitation and representation of reality. Benyus defined biomimicry as "a new science that studies nature's models and then imitates or takes inspiration from these designs and processes to solve human problems". This definition was later elaborated by the biomimicry institute (3.8) [5] to relate the biomimicry discipline to sustainability and innovation. Benyus related biomimicry to general human problems, thus extending the scope of domain applications from technology to other domains such as business, strategy, or psychology.

1.2.3.2 Biological Versus Natural Knowledge

Per definition, biomimicry is derived from the word "bio", suggesting that the knowledge is being transferred from the biological domain. However, some scholars tend to accept non animate systems such as crystals as a source of knowledge, thus extending the perception of learning to learning from nature in general, and not only from the biological domain. One could argue that nature has evolved to minimize energy consumption [8] or to satisfy other governing rules and consequently, its objects offer solutions as those available in the biological world. Practically, there is no reason to exclude any source of knowledge.

1.2.4 Book Scope

This book focuses on the biomimicry zone as a whole including the terms biomimetics, bionics, bioinspiration and BID. Whenever we use the terms *biomimetic* or *bio-inspired* design we refer to the whole biomimicry zone including BID and bionics.

1.3 Biomimetic Development Strands

Gleich et al. [4] identified three main strands of development in biomimetics. The first and oldest strand is the functional morphology, form and function. It is based on the relation of biological forms or structures and their functions. Success in this strand happens when the function is more related to its form or structure and less to the material properties. Thus, by replacing the biological material by an artificial one, we do not lose the essence of the function, as the function is derived from the structural properties and not from the material properties. Many examples of successes in this strand are connected to the field of fluid dynamics. In this case, the biomimetic design approach compensates for the limitations of mathematics and physics, by providing usable flow models that are difficult to derive mathematically. The second strand is signal and information processing, biocybernetics, sensor technology, and robotics, which is characterized by the cybernetic control loop. Examples include bio-inspired robots and biomimetic sensors. The third and recent strand is nanobiomimetics, molecular self-organization and nanotechnology. Examples include: lotus effect applications that imitate the nano structure of the epidermal protrusions and the sharkskin applications of tiles and coatings that deter bacteria.

1.4 Biomimicry Growth

The biomimicry discipline is still in formation and demonstrates high growth rates. Da Vinci Index [9] provides updated growth rates based on four area of data including—the number of scholarly articles, number of patents, number of grants and the dollar value of grants. This index is created, published, and maintained by the Fermanian Business and Economic Institute (FBEI) of Point Loma Nazarene University. This index shows clearly continuous high levels of growth. Updated index for 2015 is presented in Fig. 1.2. This growth has two main drivers: innovation and sustainability. In the next sections we elaborate on each one of these drivers.

1.5 Biomimicry as an Innovation Engine

The accelerated growth rate of the biomimicry discipline, presented in Fig. 1.2, reflects a growing number of biomimetic innovations. It is estimated that biomimicry will represent about $1.6 trillion of the world total output by 2030 [10]. Biomimicry is considered as an innovation engine that cuts across industries including traditional industries and not only the Hi-tech sector. An ecosystem based analysis of biomimicry inspired technology and product innovation, based on 218 references, revealed that material development is the largest area of biomimicry research including smart materials, surface modifications, material architectures,

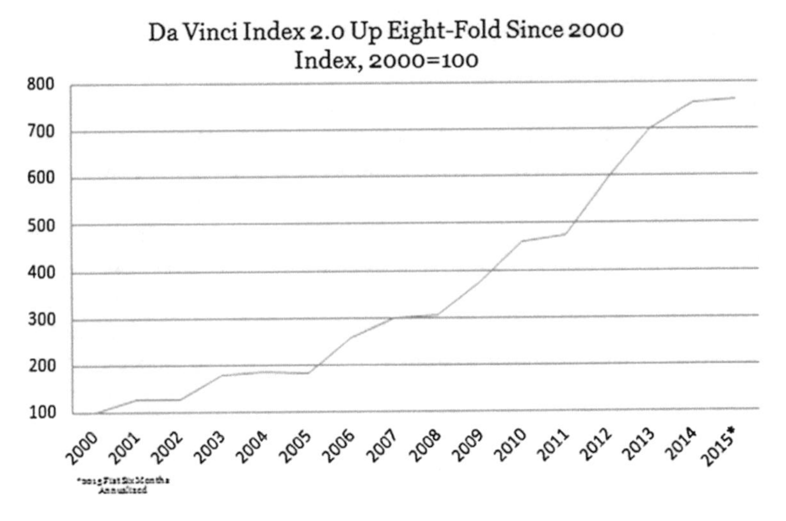

Fig. 1.2 Da Vinci index of biomimicry growth-updated to 2015

and materials with targeted applications. The second biggest area for biomimicry applications is movement applications based on animal locomotion models, including movement kinetics, release mechanisms, and structural configuration (energy efficient shapes) [11].

Biomimicry has also been identified as a mean to address the current need of sustainable innovation [12]. In this section we explore the relation of biomimicry and innovation and explain the mechanism of the biomimetic innovation.

1.5.1 Nature as an Idea Generator

Biomimicry is considered as a method for generating innovative design concepts when it is integrated during early design stages, mainly the concept design stage. During this stage, possible design concepts are generated and evaluated to choose one for detailed design.

It is known that early design stages are those that contribute most to final product quality. According to the product development paradox [13, 14] early design stages are responsible for about 70 % of the final quality of the product and the final cost, while the knowledge available to support this early design stage is scarce. The main potential of biomimicry to foster innovative solutions is therefore, when applied during early design stages when the scope of potential solutions is determined. Then, biomimicry serves mainly as a concept generator of possible design solutions. A design process with an application of biomimicry in early design stages is presented in Fig. 1.3.

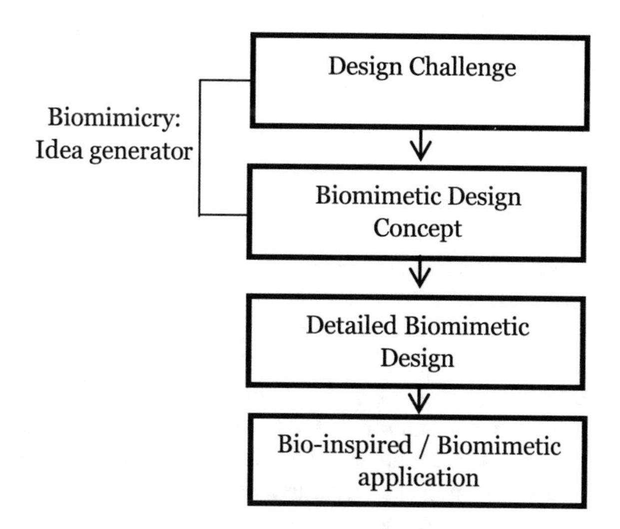

Fig. 1.3 A design process with an application of biomimicry in early design stages

1.5.2 *The Innovation Mechanism of the Biomimetic Design Process*

The innovation mechanism of the biomimetic design process may be understood by three main factors:

*More **ideas**—Distant **ideas**—Different **ideas***

1.5.2.1 More Ideas—Extending the Scope of Potential Solutions

Biomimicry extends the scope of potential solutions examined during the conceptual stages. The exact number of organisms in our world is unknown but it is estimated to be between 10 and 100 million. Only a minor number of them have been identified, and even smaller number studied. It is clear that the potential of learning from nature is huge and untapped. In addition, nature solutions are public domain. Every designer can study nature "patents" and imitate them. This is not possible when trying to use other technological solutions, because they might be patented.

1.5.2.2 Distant Ideas—Analogical Transfer Between Domains

Biomimicry by definition is based on analogical transfer of knowledge from biology to technology or other domains of application. Analogy is defined as a cognitive process of transferring knowledge from one domain of content (defined as the **source**) to another domain of content (defined as the **target**), based on similarity [15, 16]. Dealing with biomimicry, it is clear that the source domain is biology and the target domain is technology (or other domains of application) [17]. The analogy may be superficial and based on transfer of high level design principles, or extended and based on the transfer of accurate structures or processes. According to the bio-term terminology (Sect. 1.2), the first case of a superficial transfer is more related to bio-inspiration and the second case of extended transfer is more related to biomimetics.

Analogies are one of the oldest and most efficient ways to enhance innovation. Repeated findings relate analogy to innovation. Breakthrough innovations are more likely to result from far analogies between distant domains [18]. Between-domain analogies yielded more original ideas than within-domain analogies, which yielded more ideas in quantity [19]. Based on these findings, analogies used in biomimetic design, are far and between fields, and therefore have high potential of innovation. Indeed, class studies in bio-inspired design found that innovation improved by analogical reasoning that link biological functions to engineering challenges [20].

Johansson [21] referred to a similar phenomenon of bringing concepts from one field to another, and identified this process as a source of explosion of ideas and disruptive innovation. He called this process the Medici effect, a term coined after the Medici Italian banking family, who promoted innovation unintendedly, by supporting professionals from various disciplines, during the 14th century. Johansson highlighted the intersection of diverse disciplines, industries or cultures as a source of innovation. Among the rich examples Johansson offered in his book, there are also biomimetic examples such as the East Gate center in Zimbabwe [22]. Biomimetic design processes are a good example of innovation emerging from the intersection of different domains, biology and technology.

1.5.2.3 Different Ideas—Paradigm Shift

Not only nature solutions are distant from technology, but they are also based on a different paradigm. This difference is demonstrated when we compare design solutions in biology and technology, by the assistance of the TRIZ.

TRIZ is an acronym in Russian known in English as "theory of inventive problem-solving". This theory is derived from the study and modeling of hundreds of thousands of technological patents. TRIZ contains extended knowledge about the structure and nature of technological systems including the contradiction matrix that suggests 40 inventive principles to solve contradictions in technological systems, laws of technical system evolution, system modeling tools (Su-Field) and algorithm of inventive problems solving (ARIZ).

TRIZ [23] based analysis showed that there is only 12 % similarity between the principles of solutions in biology and technology. While in technology usually energy and materials are being used to solve problems, in biology solutions are based on information and structures [24].

Engineers manufacture according to a designed plan while nature grows the material together with the whole organism according to self-assembly principles [25]. Biological systems usually follow the principle of multifunctional design.

Each component has several functions, offering an elegant and cost effective design. Technological systems not always follow this principle; in many cases each component has one or only few functions and furthermore, in one design theory—Axiomatic Design (AD)—it is recommended to decouple functions [26]. In technology, more than several hundred different polymers, metals and newly developed materials are being used to manufacture products. In nature, biological materials are made of few polymers including proteins and polysaccharides. Different design principles are found in nature. For example, strait lines and right angles are common in technology but rare in nature that is characterized with curvilinear forms. Technological structures are usually smooth whereas nature structures are usually covered with protrusions. Biological solutions are more ecosystems oriented. Technological systems are not designed so far to prefer ecosystem benefits or be decomposable. In contrast, biological systems decompose almost completely, while teeth are an exception.

These differences between nature and technology create the opportunity for designers to access different solutions, or solutions that may be unusual, surprising and "crazy". The differences between nature and technology provide the opportunity to get rid of design fixations and be exposed to brand new efficient design strategies. Thus, biomimicry not only extends the scope of solutions to distant solutions, but it also provides solutions that are based on a different paradigm and therefore have the potential to innovate.

1.6 Biomimicry as a Sustainability Engine

1.6.1 A Demand for Sustainability Tools

One of the major challenges of humanity today is to provide sustainable technologies. "Sustainable development is development that meets the needs of the present without compromising the ability of future generations to meet their own needs" [27]. While the needs today are growing due to population growth and demands for higher quality of life, the ability to supply these needs is questionable. We are still depending on oil, carbon emissions are getting higher, and 90 % of the raw material is wasted during the manufacturing processes [28]. Resources and ecosystems services are declining while their demand is increasing [29]. It is time for a real change. It is clear that this change should be inherent in the way we think, design, manufacture, consume and end the use of our products.

This required change towards sustainability may be first perceived as a business constraint. New regulations and demands require management attention and financial resources. However, it is now clear that sustainability is a key driver to innovation [30]. There are more and more case studies of companies that their quest for sustainability made them innovation leaders [31]. In fact, sustainability was identified as the 6th wave of innovation since the industrial revolution, as we nowadays have a real market need for sustainability but also critical mass of knowledge and technologies to achieve it [12].

Various sustainability tools are developed to support sustainable design. Each one is derived from different approaches to sustainability and therefore focuses on different aspects. For example, the Life Cycle Assessment tool (LCA) [32] is based on viewing environmental impacts during all stages of a product's life. This tool requires basic knowledge of materials and energy consumption that does not usually exist during the concept design stage. The Natural Step tool (SLCA) [33] is based on overview of the full scope of social and ecological sustainability at the product level. The PROSA methodology provides guidelines for Product Sustainability Assessment, derived from comprehensive product system assessment and LCA [34].

There are several examples of sustainability checklists of criteria such as the design abacus tool and the design web tool [35]. These checklists provide issues, aspects, guidelines or principles to be integrated and assessed during the design stages. Such checklists are intended to ensure that potentially significant environmental issues are identified and considered as a basis for judgmental decisions. Checklists are easy to use, but cannot cover all the requirements, and may require time to check each item separately.

Research findings support the argument that environmental parameters should be integrated during the design stages that determine most of the environmental impact [36], mainly during the concept design stage. However these early stages lack detailed information that is usually required for sustainability assessment such as material and energy consumption. The manufacturability of the product is usually not considered until after it has been designed [37]. Consequently, there is a need for new tools to address sustainability aspects of design concepts.

Biomimicry has much to contribute especially during the concept generation stage. An appropriate sustainability tool for the concept design stage may be derived from nature itself, where nature sustainability design principles are identified and gathered as a tool such as checklist.

1.6.2 Learning Sustainability from Nature

Seeking nature guidance for sustainable models and measures is reasonable and has expanded in the last years. Biological systems operate within restricted living constraints without creating waste or irreversible damage to the ecosystem. On the contrary, they enrich and sustain the ecosystems.

- Nature forms and structures provide a wide range of properties with the minimal use of material or energy.
- Nature manufacturing processes are conducted within life and therefore avoid high temperature, strong pressures or toxic materials [1].
- Nature systems demonstrate efficient flows of energy and material.
- Nature products are recyclable.

However, imitating nature solution per se, without an intention to implement nature sustainability design principles, is not a guarantee for sustainability. In fact, the relation between biomimicry and sustainability is questionable. A product may be designed based on a nature innovative mechanism but later manufactured using toxins or large amount of energy. Therefore, there is a need for a broader view that observes the product as part of the system during its different life cycle stages. In this relation, Reap et al. [38] criticized the reductionist approach to biomimicry that involves imitation of form or structure, actually claiming that sustainability cannot be found under the lens of the microscope but at higher and holistic levels of imitation.

Following this questionable relation of biomimicry and sustainability, Antony et al. [39] performed the first systematic analysis to assess the sustainability of a complex biomimetic product by PROSA methodology [34]. They compared the biomimetic product with conventional alternatives and revealed that the former meets up with the state-of-the-art conventional solutions in terms of sustainability. This is a good indication for the relation of biomimicry and sustainability. Nevertheless, the relationship between biologically inspired design and sustainability should be further investigated and reinforced by more comparative studies and case studies.

Viewing nature as a source of knowledge in general and sustainable knowledge in particular encourages a paradigm shift towards a sustainable relationship with nature that is based on learning and respect, and is not based only on resource consumption.

1.6.3 Biomimicry as a Sustainability Tool

As mentioned, there are various sustainability tools, each of which is representing a different perspective. Faludi suggested to classify sustainability tools into one or more of these basic categories according to their purpose [40]:

- Suggesting specific design ideas (strategies)—guide the designer in *what* to do.
- Setting priorities/focusing attention (objectives)—clarify *what* are the design objectives and *where* to put the attention and budget.
- Keeping score (metrics)—help to measure *whether* the objectives are met.

All the three are needed for an effective design process.

Faludi elaborated on the biomimicry position in green design and suggested that biomimicry belongs mainly to the first category as it is a concept generator tool that creates sustainable innovative design ideas. It can also be used as a metric but on the meta-level, suggesting what to measure and not how to measure. For example, the idea that a city should function like an ecosystem can be derived from biomimicry, but the actual measurement of the city ecosystem services should be performed by a different tool, such as LCA. Thus, biomimicry can be integrated with other sustainability tools, especially with tools for setting objectives and metrics that can be used for deciding which of the biomimetic ideas should be further elaborated.

1.6.3.1 Approaches of Biomimetic Sustainability Tools

Nature sustainability strategies and design principle were studied, identified and made accessible to designers. This core of knowledge is referred to as the "life principles". Various attempts have been done to define life principles [1, 5, 41–43]. The core knowledge is summarized in the life principles framework of biomimicry 3.8 (Fig. 1.4).

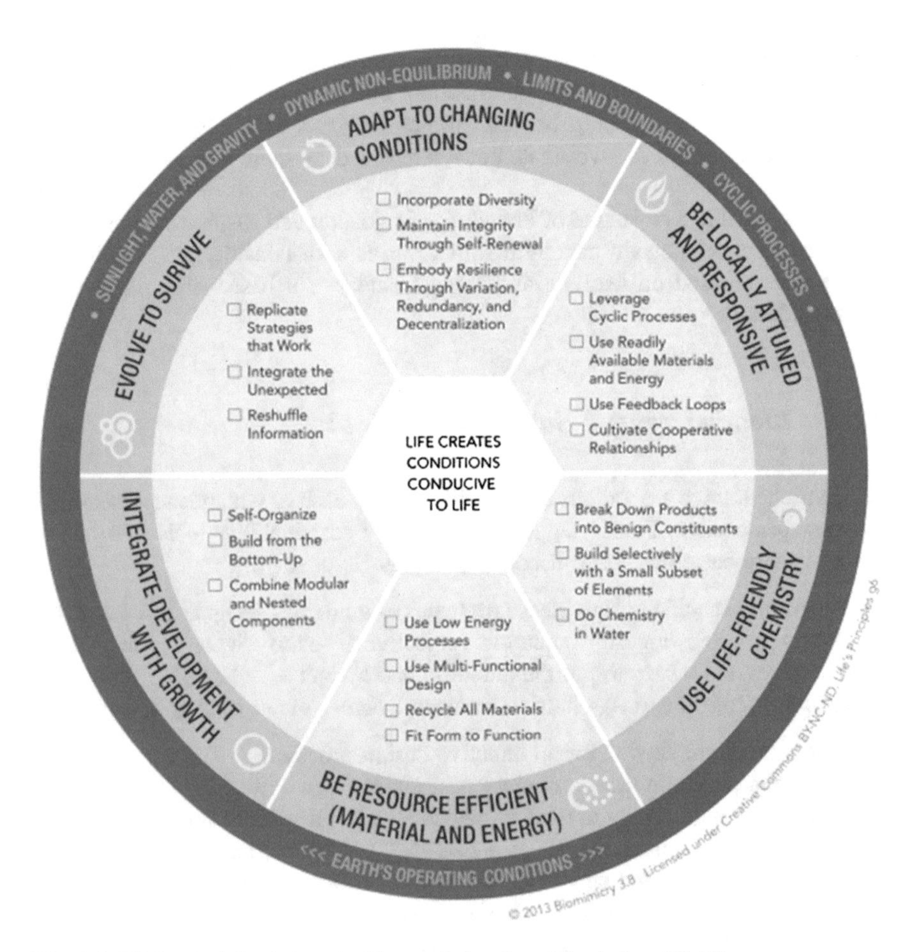

Fig. 1.4 Life's principles (adapted with permission from Biomimicry 3.8 [5])

Life principles are nature strategies and design principles that repeat among various organisms at multiple scales, and represent nature design solutions to survive under earth operating conditions, limits and boundaries. These repeating solutions are actually sustainability patterns that may be transferred to biomimetic applications. Sustainability design patterns are considered as fundamental units of the analogical transfer of knowledge from biology to technology or other domains of application [44]. According to Faludi's categories (Sect. 1.6.3), life principles provide specific design ideas (strategies) and metrics to measure whether the proposed design indeed meets nature sustainability principles.

Designers who implement these life principles during the design process may foster sustainability. However, some of the principles are general and their application in engineering is neither clear nor straightforward. In addition, it is unclear how the life principles were revealed and how to search for new ones. There is a need for a framework that could lead the search for more sustainability principles in nature and formulate them in an applicable way for designers.

Cradle to cradle [45] is another example of biomimetic sustainability tool, inspired by cyclic flow of energy and materials in nature. According to this design approach, liner systems that extract nature resources and end their life in the landfill (cradle to grave) are replaced with cyclic systems (cradle to cradle), that are designed to reassure their circulation of organic or technical materials, just like in nature. According to Faludi's categories, the cradle to cradle approach provides specific strategy of cyclic design and provides a metric to measure whether the proposed design indeed enables the biological and technical metabolisms.

Nature Inspired Design Handbook (NID) [46] offers another bioinspired design method. NID combines the life principles and the cradle to cradle design principles and offers 6 principles derived from the best design practices observed in successful living systems. These principles can be applied by using the NID wheel, a representation of eight disciplines important for the application of NID and eight elements they collaborate on. For example, design students developing a NID package for dishwashing products, offered soap enriched with nutrients for the benefit of microorganism in soil. They applied NID principles of 'waste equals food' and 'be locally attuned and responsive' by aiming to create positive impacts on the context of use [46].

1.7 The Imperfection of Nature

Some biomimicry proponents proclaim that nature solutions are the result of 3.8 billion years of evolution, trial and error field tests of nature design lab. Comparing to technological evolution period, nature design lab has an unsurpassed advantage. But we acknowledge that nature design solutions are not always optimal, ideal, elegant or perfect. Can we study also from non-ideal solutions? Miller [47] discussed the idea of design flaws in nature and suggested that complex biological systems may demonstrate errors that no intelligent designer would have performed.

One example for this imperfection is the structure of the human eye. Although the eye is a complicated designed structure and generally functions well, one design flaw may be identified. The photoreceptor cells of the human eye are located in the retina and pass neural impulses to the cells of the optic nerve through a series of interconnecting cells. The neural connection is placed **in front** of the photoreceptors cells and blocks the light from reaching them. In addition, network of blood vessels that supply blood to the nerve cells are also located directly in front of the photoreceptor cells, demonstrating another disturbance for the light sensitive layer. And finally, the neural wiring must be passed through the retina to connect the nerve impulses to the brain. The result is the blind spot in the retina. Obviously, a better place to locate the nerve cells, for example, is **behind** the retina, preventing the disturbance to the photoreceptors cells and preventing the blind spot.

A designer who starts from scratch may design optimal systems, but evolution is an incremental design process, applying minor changes to existing structures. The retina evolved as a modification of the outer layer of the brain, and light sensitivity capabilities evolved later including the nerve connections. The result is a modification of preexisting structure rather than an elegant perfect design, free of flaws.

Imperfections and designs flaws in nature can also be an innovation stimulus if we incorporate judgmental thinking and ask how we can improve the non-optimal design. Designers are the last decision makers and should consider nature solutions and understand their imperfections. Designers should consider which parts of the biological solutions are adapted, which parts should be changed, and how to implement the design concept in an elegant and sustainable way. Biomimetic design tools assist the design process but they are not a guarantee for success, just like other creativity tools [48]. Designers have the responsibility to filter nature knowledge and adjust it to the engineering design space. They should bear in mind that nature is a robust evidence for working principles. These principles are not always optimal, but they work in some certain conditions.

1.8 Biomimicry—Promises and Obstacles

The promise of the biomimicry discipline is huge. Millions of solutions are waiting to be studied. These solutions have passed the evolution test and have no patents limitation. They are based on a different paradigm, can be surprising, release engineering fixations and above all, foster sustainable innovation. Even if these solutions are not always optimal, they may serve as a source of knowledge for the design space, as discussed in Sect. 1.7.

Biological data growth is huge and basic mechanisms have been discovered, but life sciences have come to a turning point when infusion of rigorous knowledge such as engineering knowledge could sustain its further development [49]. Biomimicry provides the opportunity to make this biological data applicable and productive. If biomimicry is an innovation engine, the biological data is the fuel that activates this engine toward real innovations. However, there are some obstacles for the success of the biomimicry design process:

1. Scalability—Moving from micro to macro scale exposes designs to new constraints. Some biological mechanisms work at the nano scale but fail to work on macro scale. Example is the gecko attachment mechanism. Synthetic mimics of the gecko's attachment system have failed to show adhesive performance at large scales [50]. This is not a surprise, it happens also in technological solutions. For example, in the transfer from the macro to the micro, with the development of MEMS (micro-electro mechanical system), it became clear that friction acts very different in these two scales [51].
2. Material constraints—Sometimes there is no artificial substitute for the biological material. Mainly when the function is more related to the given material and less to the structure. One example is the spider silk. Although its molecular structure is known, the scientific world struggles to synthesize artificial materials that could imitate the structure and maintain its unique characteristics.
3. Manufacturing constrains—Manufacturing or technical issues are one of the major restrictions today on delivering biomimetic innovations. For example, the lotus leaf artificial products are still far from the biological model performance [4].
4. Irrelevant design constrains—Nature constrains are not always relevant to the engineering domain. Designer may start from scratch while nature design lab is based on incremental changes. Complexity in nature is not always relevant to engineering domains. For example, a living body needs to be adjusted to growth processes, while this constraint is not usually relevant for technological systems, though it may be a source of innovation.

Chapter 2
The Biomimicry Design Process: Characteristics, Stages and Main Challenge

There exist many perspectives about design processes in general [52]. The introduction of biomimicry adds to this complexity and variety. In order to understand this setting, this chapter describes the characteristics of the biomimicry design process, its stages and main challenges. It becomes clear that one significant issue in realizing the process is the language gap between engineering and biology.

2.1 Characteristics of the Biomimicry Design Process

2.1.1 Bidirectional Design Process

Two biomimetic design processes are identified according to their starting point. We can start from technology or engineering with a design problem and locate a solution in biology (from a problem to biology). An example is the bullet train in Japan that was redesigned after the beak of the kingfisher to solve the noise problem. First, the problem was identified as the noise the train produced every time it came out of a tunnel, due to a change in air pressure. Then, a solution in biology was found in the kingfisher which dives from air into water with little splashing. The front of the train was redesigned using the beak of the kingfisher as a model. The result is a quieter, faster and more energy efficient train [5]. However, the kingfisher was found coincidentally and not by a search process that is usually required to support a biomimetic design process from a problem to biology.

We can also start from biology with a biological solution and move to locate an application for an analogical problem (from biology to an application). One example follows the discovery of the self-cleaning mechanism of the lotus, which is based on small epidermis protrusions causing the droplet to collect pollutants while it rolls off the leaf. Several applications were found for this biological effect including a self-cleaning paint, glass and fabric [53].

Y.H. Cohen and Y. Reich, *Biomimetic Design Method for Innovation and Sustainability*, DOI 10.1007/978-3-319-33997-9_2

Table 2.1 Biomimetic design directions terminology

	From biology to an application	From a problem to biology	Source	Year
1	Solution driven	Problem driven	Helms et al. [54]	2009
2	Organism driven	Mechanism driven	Hesselberg [55]	2007
3	Bottom up	Top down	Speck and Speck [56]	2008
4	Biology push	Technology pull	ISO Biomimetic committee [57]	2012
5	Biomimetics by induction	Biomimetics by analogy	Gebeshuber and Drack [58]	2008
6	Biology to design	Challenge to biology	Biomimicry 3.8 [5]	Unknown

Helms et al. [54] identified these bidirectional design processes as "Problem-driven" and "Solution-driven" biologically inspired design. Other researchers called these directions with different terminology as presented in Table 2.1. Whether we start from biology and end with technology or vice versa, at the end, knowledge is being transferred from biology to technology to solve technological problem, generate a new capability that may solve the problem, or replace other inferior solutions. In the first case, interesting biological phenomenon sparked the process, and in the second case, a technological need sparked the process.

Each one of these design directions is a biomimetic design process, while the direction "from biology to an application" is more common according to our research database described in Sect. 5.4. It might be easier to find analogical design problems to a given biological solution then finding an analogical biological model to a given problem among the millions of potential biological sources.

2.1.2 Analogical Based Design Process

Biomimicry by definition is an analogical transfer of design knowledge between biology (the source) and technology (target), or others domains of applications [17]. Biologically inspired design often involves compound analogies when a design concept is generated by multiple cross-domain analogies [59]. In this case, different organisms may be sources for different functions that at the end will be integrated in one technological system. This integrative approach [55] was demonstrated by a case of developing an endoscope inspired by rag worms which can move on slippery surfaces.

2.1.3 Interdisciplinary and Multidisciplinary Design Process

Mutidisciplinary study involves different disciplines, where each one provides a different perspective on the subject under study, but these perspectives are not integrated. Interdisciplinary study on the other hand, involves an integration of theoretical, conceptual and methodological approaches of different disciplines [60]. Thus, an interdisciplinary process requires more flexibility and blurring of boundaries so a new body of knowledge can emerge.

Every design process incorporates multidisciplinary teams involving at least marketing people, engineers, and customers. Biomimetic design involves also biologists thus the level of complexity is even greater. However, biomimetic design process goes beyond the level of viewing the design challenge from different perspectives (multidisplinary), as it involves also an integration of disciplinary knowledge (interdisciplinary). It is clear that each discipline benefits from this interdisciplinary process. Studying a biological system in the context of biomimetic design may extend biologist's understanding of an observed mechanism. For example, Full [61] studied the gecko's movement and evoked interest about the role of the tail. He provided biologists with a hypothesis about the role of the tail, one they did not considered. Thus, collective discoveries emerged beyond any single field. Full called this mutual benefit for both disciplines 'Biomutualism'. Reich et al. [62, 63] called the process where multiple disciplines are integrated and enrich each other over several cycles a bootstrapping effect.

Parvan et al. [64] offered a collaboration model between biologists and engineers referring to their tasks and roles in the process. Schmidt [65] suggested that biomimetic knowledge starts from a zone of communication between an engineer and a biologist, where a circulation of knowledge occurs rather than a unidirectional knowledge transfer. This circulation of knowledge, leading also to knowledge extension, is again a form of bootstrapping. Another collaboration model developed by Swedish biomimetics 3000 [66], the $V^2IO^{®}$ model, is said to be an innovation accelerating model that integrates multiple disciplines, organizational issues, cultures and their corresponding global challenges. However, its details are not disclosed, hence it cannot be assessed. Nevertheless, the idea of creating a model that accounts for the impact of the social and organizational aspects to the innovation process is valuable. PSI—product, social, institutional—is a general framework that considers all these issues [67–69].

2.2 Biomimetic Design Process Stages—From a Problem to Biology

Figure 2.1 describes the biomimetic design stages from a problem to biology. When we start from a problem we first need to define the problem (stage 1). This stage is not unique to biomimetic design as every design process starts with a problem definition task; a good definition of a problem is considered to be a major part of the solution. However, in the context of biomimetic design a problem definition has even a greater impact of bridging to biology (stage 2). After the problem definition, we follow with three core stages of every biomimetic design process that are unique

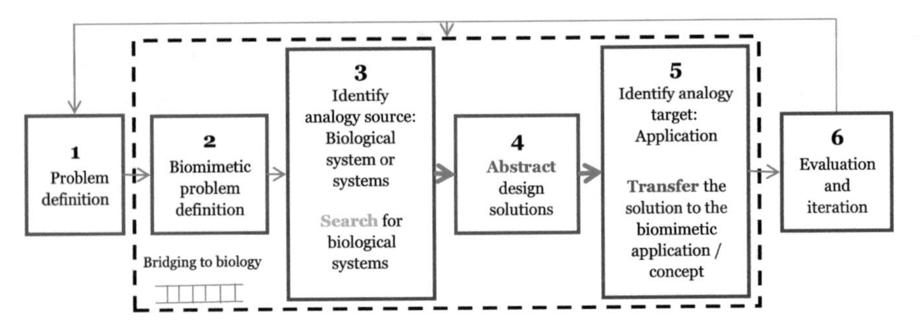

Fig. 2.1 The biomimetic design process—from a problem to biology

to biomimetic design and are derived from the biomimetic analogy: Identify analogy source—the biological system or systems in case of compound analogy (stage 3), abstraction of the biological solution (stage 4), and transfer the solution to the application (stage 5). These three core stages were also suggested as a base to define a biomimetic product, as part of the ISO standardization initiative that is based on VDI Guideline 6220 [70].

At the end of these three core stages, there is an evaluation and iteration stage (stage 6). This stage is also not unique to biomimetic design, and characterizes every design process. A designer may want to repeat each one of the previous stages, including the problem definition, location of the biological system, abstraction, or transfer to biomimetic concept or application. The dashed box in Fig. 2.1 includes the stages unique to biomimetic design. Outside the dashed box are the initial and final general design stages of problem definition and evaluation. The bolded arrows between the source (biology) and target (application) represent the connecting bridge between these two domains.

2.2.1 Problem Definition (Stages 1 and 2)

Problem definition is derived from customer needs or from some observed opportunity and transformed by designers to a technical definition that drives the design process. The result of this stage could be several problem definitions according to different technical views or different interpretations of the initial needs. In order to search for relevant solutions in nature, we need to move from a technical definition to a biomimetic oriented definition that supports the biological search. We call this process bridging to biology as we set here the foundation for the bridge that is being built later during the abstraction (stage 4). Therefore, problem definition in biomimetic design process from a problem to biology has an implicit role in the retrieval process of biological systems.

Biomimicry 3.8 [5] offered to 'biologize' the problem by redefining problems with biological terms. Helms [54] suggested to reframe the problem by biological terms and ask "How do biological solutions accomplish xyz function?". Other problem definition approaches directed designers to define clearly the function of interest, assuming that a clear function could lead the search process and clarify the design target. Sartori et al. [71] provided guidelines for this problem definition/analysis phase: (i) identify the required function from problem description and (ii) identify the most important requirements and conditions. All these suggestions are clear but too general and opened to personal interpretations without tools to 'biologize' the problem.

Mckeag [72] did offer a tool, the Bio-design cube to accomplish the translation to biology. Each side of a cube represents a problem definition space: What it is (form, process, system), what is the key parameter (information, energy, structure), and where it can be applied (Science, Design, Business). All the aforementioned ideas and tool are merely proposals for bridging. They need to be further investigated in relation to biomimetic problem definition to assess their value.

2.2.2 *Identify the Analogy Source: Search for Biological System (Stage 3)*

In a biomimetic design process from a problem to biology, we first need to perform a retrieval process in order to find a biological system that demonstrates a required solution for an analogical problem. This process requires searching algorithms and techniques to retrieve relevant information from biological databases. Elaboration on this issue is presented in Sect. 3.1. The result of this searching process is an identification of a biological system. The identification occurs when the designer acknowledges the analogical relation of the biological system to the given problem.

2.2.3 *Abstraction—Abstract Design Solutions (Stage 4)*

Abstraction in the context of biomimetic design is the process of refining the biological knowledge (design solutions) to some working principles, strategies or representative models that explain the biological solution and could be further transferred to the target application. It may be understood also by the word 'Simplify' the biological complexity into some transferable design mechanisms or principles. During the abstraction stage, the bridge between biology and technology is built and the biological system is presented in the context of analogical reasoning. This bridge creates the language that allows a designer to go back and forth, detaching from one domain and moving to the other to transfer the required knowledge. Therefore, the abstraction stage is the core of the biomimetic design process.

Abstraction is considered to be one of the most difficult steps in the biomimetic design process [56]. In fact, the transfer of knowledge is done from a model of a biological system to a model of a technological system [71] so during the abstraction stage, we aim to create a model of the biological solution. This model should explain how the problem is solved in biology and may include references to functions, structures, behaviours, design principles or strategies in case they are related to the solution. Indeed, abstraction often involves representation of the biological solutions by models. Representation of knowledge is a natural cognitive process executed when trying to understand a phenomenon. In relation to biological phenomena, representations can facilitate understanding the functional mechanism and support the transfer to technology [73, 74]. Vattam et al. [75] conducted a class experiment in a course of bio-inspired design. Most of the 45 students were senior students from engineering background. They found out that the students used rich mental representations at different levels during the biomimetic design process, and suggested that it had an advantage in creating biomimetic analogies.

Mak et al. [17] demonstrated the importance of abstraction to the sequential transfer stage of biomimetic concept. They found out that abstracted principles regarding biological solutions tend to evoke more biomimetic concepts, comparing to information about forms and behaviours of biological systems. However, the advantage in quality of these concepts was not discussed extensively. They also reported on difficulties with the analogical mapping due to superficial analogies and fixation problems when a designer tends to stick to the first biological solution he encounters [76].

Abstraction requires knowledge about the biological solution in order to model it. Therefore, the biological system should be first analysed and understood. However, the available knowledge in the literature might not be sufficient to understand the biological mechanism. In this case, there is a need for further investigation.

2.2.4 Transfer the Solution (Stage 5)

Following the abstraction stage, the abstracted knowledge should be transferred to technological or other domains of applications. We transfer the knowledge that is relevant to the solution we aim to imitate. There are different levels of knowledge transfer including forms, structures, processes, functions, systems or principles [77].

Schmidt [65] identified three levels of knowledge transfer including (i) Structures, forms and materials (ii) Functions and (iii) Processes and information. Sartori et al. [71] identified four levels of knowledge transfer from the biological to the technological system, based on the SAPPhIRE model: (i) Parts—same materials in the same arrangement; (ii) Organs—same or similar organs including the physical effects related to these organs; (iii) Attributes—same attributes (properties of the parts); and (iv) State of change—the state change of a biological system is transferred but implemented with technical means, without using the same organs or physical effects like in biology. Sartori et al. [71] reported that the transfer most often was carried out at the physical effect level that is more related to the organs of the system, i.e., the system structures. Jacobs et al. [78] reached a similar conclusion regarding the question of what is being transferred from biology to the applications. They performed a quantitative and qualitative analysis of the BioM database of biomimetic innovations and found out that the majority (61.8 %) of biomimetic designs incorporate elements of biological form, most of which (51.3 %) include only Form.

The knowledge may be extracted from different levels of organization of living things, from the cell, to organs, organisms and ecosystems. There are many examples of transfer of forms and structures such as biomimetic materials, coatings, adhesives and functional structures. There are also examples of transfer of processes, such as genetic algorithms and swarm intelligence algorithms.

2.2.5 *Evaluation and Iteration (Stage 6)*

Following the transfer stage, a designer should evaluate the results and repeat the process again if required. The initial biomimetic design concepts may be abandoned later due to various reasons such as technological or manufacturing constrains, costs and complexity. In this case, a designer may go back to nature to generate new design concepts that will hopefully meet the expectations this time.

Biomimetic design process is not linear but iterative. This iterative nature is expressed by the biomimetic spiral diagram of biomimicry 3.8 [5], that integrates feedback loops within the design process. Gramann [79] offered three evaluation steps that lead to either transfer to biomimetic application or repetition to a previous stage in the process. (i) Evaluate the analogy—the initial analogy might be superficial, inaccurate, or lead to a dead end. For example, when InterfaceFLOR® searched for substitutes to replace the glue they used to fix the modular carpets to the floor, they first searched analogies for glues in nature. This level of analogy led them to a dead end as they observed complicated adhesion mechanisms that required significant R&D effort. When they rephrased the analogy and searched for principles to stay attached in nature they extracted the principle of using gravity for attachment. That led to their novel TacTiles technology that adhere modular carpet tiles to each other rather than to the floor, and let gravity holds them in place [31]. (ii) Evaluate the abstracted model—the abstracted model may be short in description and we may want to consider adding more aspects. For example, if we initially based the model on the organism itself, we may want to include its interaction with other environmental elements. (iii) Evaluate the organism—we may conclude that the organism search was not accurate and define a new objective search that would lead to new biological models. This iteration is important because sometimes the initial biological system being observed is not the most suitable solution. A tendency to fixate on the initial biological phenomena was observed by Helms et al. [54] in a research conducted on students. Iteration may address this fixation but other means might be necessary.

2.3 Biomimetic Design Process Stages—From Biology to an Application

Figure 2.2 describes the biomimetic design stages from biology to an application. When we start from biology and move to the application, we have similar core stages with slight modifications. First, we encounter a biological system with a unique characteristic or mechanism (stage 1) and we identify it as a suitable analogy source when we realize its potential benefit to innovate. When we encounter several biological systems with unique characteristics, we may form compound analogy.

Many biomimetic innovations sparked in a moment of wonder from a biological mechanism. The lotus effect was discovered after the wonder of observing a clean lotus leaf in a dirty environment. The mystery of how penguins stay ice-free though they live in very cold temperatures led to a biomimetic research to prevent ice formation on airplane wings [80]. The fast process of building a new crayfish skeleton in freshwater environment aroused the curiosity of a crayfish farmer. The research led to a discovery of an Amorphous Calcium Carbonate (ACC), a base of a new bioinspired calcium supplement [81]. Metaphorically, we may call this moment the Bio-WOW (stage 1), the wonder stage that sparks the biomimetic innovation process.

Following the Bio-WOW stage, we define which problem is actually being solved at this biological system that sparked our wonder (stage 2), in order to identify analogical challenges in technology and possible applications (stage 3). At this stage we do not choose the specific biomimetic application but we do acknowledge the relation to some possible applications. This relation motivates us to keep on with the analysis and abstraction process to gain better understanding of the biological solution. Therefore, stages 1–3 are interrelated as we do not proceed with the process of analysing and abstracting the biological solution, unless we respect its value to spark innovation. Compared to a biomimetic design process that

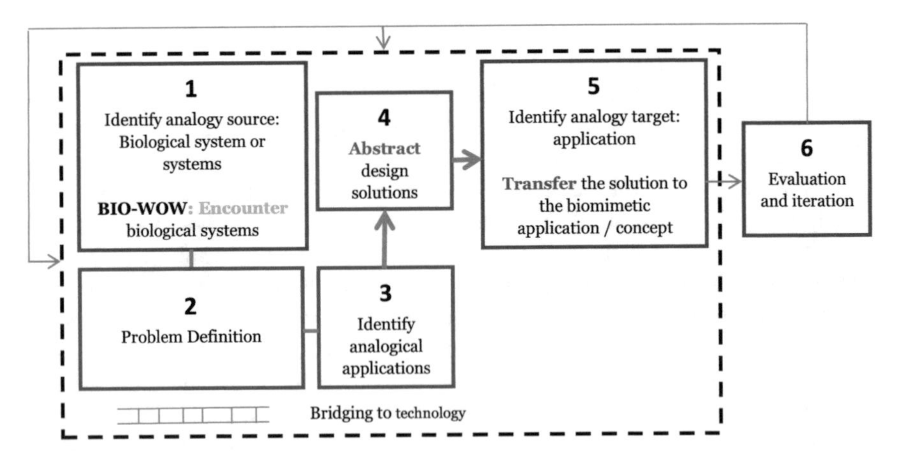

Fig. 2.2 The biomimetic design process—from biology to an application

starts with a problem (2.2), here we do not need to perform a retrieval process as we start from the biological system we encountered (stage 1). Next, we move to build the bridge to the application by the abstraction stage (stage 4) that enables to transfer design solutions between the domains. Then, we transfer the biological solution to a suggested biomimetic concept or application (stage 5). At this stage, we have deeper understanding of the biological solution and we are able to define a specific application. Finally, we evaluate and perform iteration to each one of the previous stages if required (stage 6). Elaboration on the abstraction, transfer and evaluation stages (4–6) in Sect. 2.2 is also relevant to this design direction.

The dashed box in Fig. 2.2 includes the core and unique stages to biomimetic design. Outside the dashed box is the final general design stage of evaluation. The bolded arrows between the source (biology) and target (application) represents the connecting bridge between these two domains.

2.4 The Synapse Design Model Charts

The above mentioned design processes from a problem to biology and vice versa, are presented in a more appealing way in the two following synapse design model charts. Synapse, the space between nerve cells where signals are passed, is named after the Greek word synapsis that means a conjunction or point of contact. The following biomimetic design model charts are named after the synapse conjunction concept, as they aim to connect biology and engineering (see Figs. 2.3 and 2.4).

The synapse design charts steps, one by one, lead the designer in the conjunction of biology and technology. As the model suggests, iterations forward and backward may be required during the process.

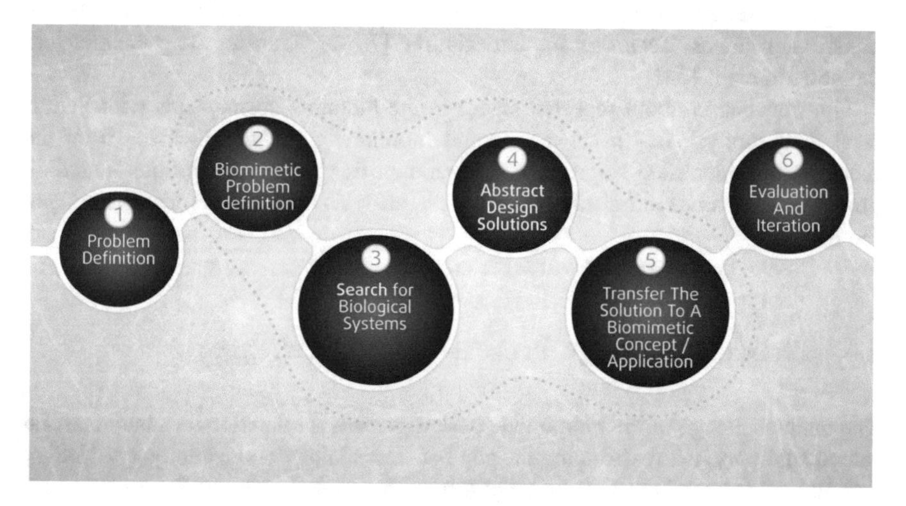

Fig. 2.3 The synapse design chart: from a problem to biology

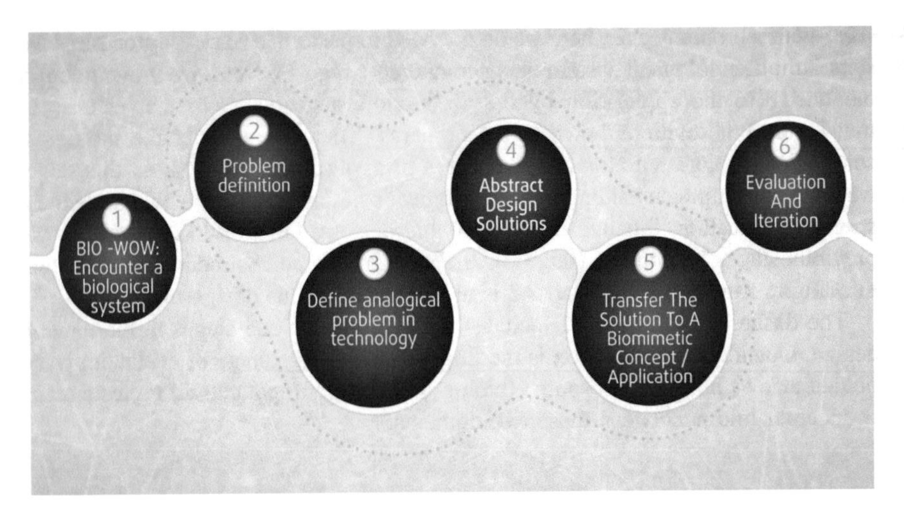

Fig. 2.4 The synapse design chart: from biology to an application

2.5 Biomimetic Design Process Stages—Literature Review

The stages we described in Sects. 2.2 and 2.3 as our suggested model for the biomimetic design process appear in the literature in various sources. For example, the stage of biological system search appears as 'Biological solution search' [54] or 'Search for examples of relevant biological systems' [79] for a problem to biology direction. The stage of encountering a biological system appears with partial similarity as 'Biological solution identification' [54] for a biology to application direction. The stage of abstraction appears as 'Identify solution principle in the biological example' [71] or 'Analyze the biological systems' [79]. The stage of transfer appears as 'Technical implementation' [79] or 'Transfer the principles into the new domain' [54].

Our conclusion about repeated stages in the literature corresponds with Sartori et al. [71] study. They reviewed several biomimetic design processes from the literature and identified stages that are common in various descriptions including problem definition, biological system search, analysis of the biological system and the transfer.

2.6 Biomimetic Design Process—Main Challenge

The main challenge of the biomimetic design process is related to its analogical and interdisciplinary nature: bridging the gap between biology and technology. Biology and technology are based on different terminologies and ways of thinking.

In general, knowledge structure and mapping in different disciplines is different [82], although there is notable exception, namely, the Interdisciplinary Engineering Knowledge Genome (IEKG) that bridges between seemingly distinct disciplines [62, 83, 84]. The difference between biology and engineering was previously discussed. In addition, functional terminologies in both domains may be different. For example, the function 'Transport' may have different meaning in biology. While plants do not move they can be a source of inspiration for transportation by exploring the way they disperse their seeds [85]. We elaborate on functional terminologies difference in Sect. 6.1.2.

The difference between biology and technology is also manifested by the way engineers and biologists are trained and think. Biologists are trained to investigate phenomena and search for details. They do not tend to abstract or formulate rules. Engineers are trained to solve problems by synthesis and they investigate possible solutions. In general, biologists lack engineering and design knowledge and engineers lack biological knowledge. The main challenge therefore is bridging these two domains and creating an infrastructure for fertile knowledge transfer and communication.

Chapter 3
Biomimetic Design Methods—Literature Review

Generally, we may say that the biomimetic design stages described in Sects. 2.2 and 2.3 are intuitive. It is clear that if we want to transfer knowledge between domains, we need to define the problem, search for biological solutions, identify core design principles and transfer these principles to technology or other domains of applications.

However, these intuitive stages do not provide an answer for the main challenge of the biomimetic design process: bridging the gap between biology and technology. They mainly answer the question of "What" should be done in the biomimetic design process and not "How" to do it. These stages do not provide tools to support the core stages of biomimetic design:

- Biological system search—How to find in nature suitable systems for inspiration or imitation?
- Abstraction—How to model biological cases and analyse their models with relevant analogies for applications?
- Transfer—How and what to transfer to the domain of application?

Several attempts to answer these main questions can be identified in the literature and accordingly, they fall under three high-level categories:

i. Searching/retrieval methods—Methods to support search and retrieval of biological solutions.
ii. Abstraction Methods—Methods and tools for analysis and representation of biological systems as a base for analogical reasoning.
iii. Transfer methods—Guidelines to assist designers during the transfer process.

Similar categories of biomimetic design methods appear in the literature [2]. It is clear that the three categories are required to complete a biomimetic design process. Abstraction cannot be performed without locating a biological system, and transfer to application cannot be completed without the abstraction stage.

Y.H. Cohen and Y. Reich, *Biomimetic Design Method for Innovation and Sustainability*, DOI 10.1007/978-3-319-33997-9_3

In the following sections we summarize the main efforts of each category based on a literature review. These efforts support the creation of a biomimetic methodology. So far, the literature is abundant with biomimetic examples and case studies but the field still lacks a theoretical and applicative framework that could bridge the gap between biology and technology. Though the efforts described in this chapter support the formation of the biomimicry discipline, they are still more based on description of ideas and less on evidence. Most of the efforts we describe in this chapter are evaluated by lab experiments performed in an artificial environment of academic exercises, mainly in classes. Obviously, the conclusions are limited and should be strengthened in the future by field experiments addressing real design challenges. Moving towards a stage of demonstration and evaluation is mandatory for the formation of the biomimicry discipline.

3.1 Searching/Retrieval Methods

Finding a biological system among the millions of possibilities is a difficult task. Above the huge quantity of possibilities, engineers usually lack biological knowledge that might be required for this searching process. In addition, the analogical solution may be found in several levels of biological organization, from the molecular to the ecosystem/biosphere levels [2], so the searching process is even more complicated and time consuming. Studies in lab experiments showed that up to 25 % of out-of-class time may be spent on searching for the biological organism [86]; It is reasonable to assume that in reality the searching process would be much longer.

In some biomimetic innovations it is not clear how the biological system was found, or it is clear that it was found by accident. For example, the engineer of the bullet train in Japan was lucky enough to watch a kingfisher diving from air into water with little splashing. He realized that the kingfisher actually copes with a challenge of passing from one medium to another without noise [5]. What would have been the future of the bullet train if the kingfisher did not appear that day in front of the engineer? We must explain how to position the kingfisher as a model out of the millions of potential models in a systematic way and not by chance, or how to find other biological sources that may even improve the present design. In this section, we summarize the major approaches, guidelines and tools to perform the biological system search.

Three levels of search are identified: (i) Consulting individuals; (ii) Searching designated biomimetic databases; and (iii) Searching general biological sources. It is clear that the amount of knowledge investigated increases from level to level, but each level has its own advantages and disadvantages.

3.1.1 Consult Biologists

Intuitively, biologists who are familiar with biological systems are a good place to start. Engineers who are usually not familiar with the biological domain can partner with biologists as an access gate to biology. However, biologists are not always willing to cooperate, and if they are, their knowledge is limited compared to the huge potential of nature. They also might be biased by their areas of expertise, and not necessarily be able to recall objectively potential analogies [2].

3.1.2 Search Designated Biomimetic Databases

There are two types of biomimetic databases: (i) databases of biological systems that are classified according to engineering terms or other relevant taxonomies and (ii) databases of bio-inspired products. The first type is a good source to locate biological organisms. The second type is a good source to locate biomimetic applications; however these applications are related to biological systems that may be used to inspire new biomimetic innovations.

3.1.2.1 Biomimetic Databases of Biological Systems

- The "AskNature" database [87], a project of the Biomimicry 3.8 institute, is an open source database that classifies biological data per functions. The database includes about 2000 biological systems. Each record includes a short description of the biological mechanism, references, images and even biomimetic application ideas. A search can be done either by functions or by the organism name. Based on the authors' experience, the user interface is friendly and convenient and the search results are relevant to the design challenge when searching by functions.
- *Biomimetics Image retrieval Platform [88]* is a project under development in Hokkaido University, Japan. Most of the searching methods are based on textual searching and keywords. In contrast, the Biomimetics Image retrieval Platform is a biological database that contains a large number of images. Instead of searching texts by keywords, one may search by using their own image data. Images are derived from biological databases of various species (insects, fish, etc.) and may be even integrated in the future with images from other fields such as material sciences and mechanical engineering.

3.1.2.2 Biomimetic Databases of Products

These databases include information regarding biomimetic applications and their biological source of inspiration. Examples of biomimetic developments may assist designers to understand how to implement biomimicry during the product development process. As far as the author knows, there are no evaluation studies of these databases.

- Bio-Inspired Products and Concepts Repository

Bruck et al. [89] created a design repository of bio-inspired products and concepts to support bio-inspired robotic projects for senior mechanical engineering students. The bio-inspired products are devices or processes in any stage of development. The repository contains 85 records of bio-inspired products. A functional description template is used to record the functions of the product or the biological system. The template includes reference to the **behavior**, the **entities** that participate in that behavior, and the **characteristics** of the behavior. A user may use free form text to search information on bio-inspired work. The products or concepts that are retrieved can then be used to design new bio-inspired products, as they provide reference to case studies.

- The BioM Innovation Database Project

The BioM Innovation Database [78] is a detailed database of biomimetic design cases. By Aug 2013, the database contained 385 cases and fifty interviews with developers. The database aims to support research on biomimetic design but can also be a source to locate bio-inspired developments.

Other biomimetic databases, such as DANE, "Idea-Inspire" and BioTRIZ, are described under the section of the abstraction methods (Sect. 3.2). They provide both an abstraction method for representing biological systems and a database to search for solutions analyzed according to the suggested abstraction method.

3.1.3 Search General Biological Databases

This searching approach is based on the assumption that a systematic search that involves logic and method yields better findings compared to unsystematic search. We present two approaches for efficient exploration of biological databases: keywords and heuristics.

3.1.3.1 Search by Keywords

Most of the searching keywords are based on functions that are described by actions. Verbs are strongly preferred over nouns as keywords to conduct searches [90] and are considered to provide a wider range of analogical biological systems [2]. Thus for example, if we search by the function "attach" we will get more analogies compared to a search by the noun "glue". The idea that a function can be the common chain that connects biology and technology is widely presented in the literature, and sometime is referred as the function bridge [5]. However, in order to search general databases by functional keywords, there is a need to connect functional terms in biology and technology through function based repertories or thesauruses. We present different approaches to identify functional keywords and related biological terms.

- A Natural Language Approach to Locating Biological Meaningful Keywords

Shu et al. [90, 91] applied a lexical language framework to systematically map engineering terms to biologically meaningful keywords. According to this approach, keywords describing engineering problems are searched by computerized searching tool with natural language format, in various biological sources, including books and papers. This search leads to identifying meaningful biological keywords for searching relevant but not obvious analogies in biological texts. Shu et al. described strategies to conduct this search and explained the relations of biological phenomena description and problem solving. It might be difficult to manage the quantity and the quality of results by this approach [2] as it may end up with too many biological meaningful keywords. Cheong et al. [92] addressed this difficulty with a refinement of this lexical approach and offered to use specific semantic relationships as criteria to identify biological meaningful keywords and also apply the lexical approach to the functional basis of Stone and Wood [93], a generic set of functions (represented by verbs) and flows (represented by nouns).

- Engineering-to-Biology Thesaurus Functional Keywords

Based on Pahl and Beitz [94] proposal to evolve a set of functions and flows to support design processes, Nagel et al. [95] presented an engineering-to-biology (E2B) thesaurus that relates biological functions and flows to engineering terms. This thesaurus contains representative set of engineering and biological terms, fosters associations between the engineering and biological lexicon, and provides several opportunities for interfacing with biological information.

Later, Nagel et al. [96] integrated this functional basis that is enhanced by the E2B thesaurus with a concept generation software model (MEMIC) and presented their computational approach to biologically inspired design. According to this approach, the software accepts a functional input and searches both a biological knowledge base (Textbook of Biology) and an engineering knowledge base (Design repository containing descriptive product information [97]). The result is more than just locating a biological system as MEMIC generates concept variants for each function–flow pair of the input. They described a case study of smart flooring to illustrate the proposed computational concept generation; however, further developments and more evaluations are required. This process provides solutions that can be implemented in design, but it is limited by the size and quality of the engineering and biology databases that could include an extended range of products/biological data [96].

- Keywords of Associations between Technical and Biological Functional Terms

Lindemann and Gramann [98] proposed a checklist of associations to relate technical functions and biology terms. They offered to link between technical functions and correlated biological associations that include biological phenomena, behaviors, or organisms. For example, the function of ‘condense gas’ is associated with leaf. Studying leaf in relation to water condensing can provide information on gas condensing. However, a complete list of the associations is not provided.

- Keywords of Relations between Contradiction Solving and Structures

Hill [99] aimed to find relations between biological structures and the contradictions they solve (based on the TRIZ framing problems as contradictions). Hill analyzed biological systems that have similar functions and their relevant structures or sub structures in order to discover the underlying principle for solving contradictions that can be transferred to technology. Associations of these structures to principles may assist the search process; however, the structures or principles are not described in details, though it is mentioned that they are stored in a catalogue.

3.1.3.2 Searching by Heuristics

Searching heuristics aim to extend or narrow the searching space in order to find more relevant results if the space is narrow or fewer results if the space is too large. Several heuristics are adapted from Helms et al. [54] and presented in Table 3.1. These search heuristics provide keywords that may be useful when searching the literature for biological systems.

"Champion adapters" is one of the most known heuristics. It was also suggested by biomimicry 3.8 [5] and by Yen et al. [100]. Extreme habitats provide survival challenges that leverage innovative design solutions. Searching for "Champion adapters" that cope with extreme habitats is a promising path for innovation.

Based on our experience as biomimetic practitioners, two additional searching heuristics may be identified:

- Search by type of zoological department and required field of innovation

An ecosystem based analysis of biomimicry inspired technology and product innovation, based on 218 references [11], clarified what types of species have

Table 3.1 List of searching heuristics

Heuristics	Description
Change constraints	If the problem is narrowly defined, such as "keeping cool", change the constraints to increase the searching space, for instance to "thermoregulation"
Champion adapters	Find an organism that survives in the most extreme case of the problem being explored. For instance, for "keeping cool", look for animals that survive in dessert or equatorial climates
Variation within a solution family	Find organism "families" that have faced and solved the same problem in slightly different ways For instance, the many variations on bat ears suggest deeper solution principles for echolocation
Multifunctionality	Find organisms or systems with single solutions that solve multiple problems simultaneously

Table 3.2 Biology and engineering: system parts analogy

	Engineering	Biology
1	System	Body
2	Casing	Skin surface, cuticle
3	Structural support	Skeleton and bones
4	Computer and control system	Brain
5	Electricity and network system	Nervous system
6	Sensors	Sensory system
7	Actuators	Muscles

inspired innovation across different areas. This understanding may be used to indicate where to search biological solutions according to the required field of innovation. For example, in material design, arthropod and plants have led to the greatest number of ideas, research and prototypes. In sensing applications, the most inspirational species are mammals. In robotics and movement applications, insects inspired the most innovations.

- Search by analogy to system parts

Bar Cohen discussed the similarities of biology and engineering systems in his book Biomimetics: Biologically inspired technologies [101]. Generally, biological terms can be described analogously to engineering terms, as both natural and artificial systems depend on the same fundamental units. Based partially on this analogy, we present in Table 3.2 basic parts of engineering systems and their analogically related parts in biological systems.

This table may be used as a searching heuristic, as we may focus only on biological parts related to the design challenge. For example, if we search for restraining solutions in nature, we may focus on skeletons and bones as they are analogical to structural support.

3.1.3.3 Defining Keywords by Heuristics: Dynamic List of Keywords

The above mentioned heuristics may be also used for the purpose of defining specific searching keywords for different design challenges. While keywords of associations between technical and biological terms, such as functional keywords, are a suggested closed list of connecting terms, here we have a method to generate different keywords for different design challenges.

For example, suppose we search for biological solutions for harvesting solar energy. Searching only by these keywords "harvesting solar energy" yields too many results in general data sources. We hereby demonstrate how to use the above mentioned searching heuristics (Sect. 3.1.3.2) in order to define relevant searching keywords, for the design challenge of "harvesting solar energy".

First, based on the "Champion adapters" heuristics, we may ask in which habitats harvesting solar energy is challenging. The answer may be three relevant searching keywords: *Deep Ocean, Poles,* and *Forest.*

By using the heuristic of variation within a solution family we may locate some "families" that harvest solar energy in different ways such as: *Corals, Plants*, and *Bacteria.*

Then, we may use the analogy to system parts heuristic and ask what system parts are relevant to a challenge of "harvest solar energy"? The answer provides us with some more keywords: *Cuticle, Leaf surface.*

Each one of these collected keywords and each keywords combination, may be now used in addition to the initial keywords of "harvest solar energy" in order to refine the searching process and access more relevant results.

This process of defining keywords by using the searching heuristics is a dynamic process that provides us with different keywords for different design challenges.

3.1.4 Searching Methods—Summary

We presented three levels of biological search including consulting biologists, searching biomimetic databases and searching general biological databases. The advantages and disadvantages of each level are presented in Table 3.3 and partially appear in the literature [2, 91, 92]. By moving from one level to another we expand the scope of possible solutions, as we have an access to an extended knowledge base.

Table 3.3 Advantages and disadvantages of searching levels

	Disadvantages	Advantages
Level 1: consult biologists	1. Search results are limited to what is known by the biologists and may be biased by their research interest 2. Dependence on biologist's cooperation	1. Communicating with a person: possibility to ask, explain and clarify
Level 2: search biomimetic databases	1. Search results are limited to what was entered into the database 2. Search results may be biased by the data categorization method 3. Need to enter data manually. Classifying biological data in such a manner presents an immense mission 4. Biomimetic databases include only a small part of the biological knowledge and the chance to innovate is reduced	1. No dependence on biologists 2. The quality of information: high chance to find relevant analogies since the data was manually checked and entered
Level 3: search general databases	1. Requires searching tools 2. The quality of results might be low compared to biomimetic databases, as the data was not entered manually	1. No dependence on biologists 2. Access to the vast biological knowledge

3.2 Abstraction Methods

The second group of methods addresses the abstraction stage of the biomimetic design process. During this stage, a designer needs to gain sufficient understanding of the biological phenomena that allows him to refine and extract design principles in the context of analogical reasoning. Further elaboration on abstraction and representation models was presented in Sect. 2.2.3.

In the following section we introduce several representation models that are used during the abstraction stage of the biomimetic design process. Some of these models are accompanied with databases or computational tools that assist the modeling process and provide examples of biological solutions analyzed per this model.

3.2.1 Abstraction Methods Without Databases

3.2.1.1 Functional Modeling

Functional model (also referred as functional analysis or functional decomposition) is a "description of a product or process in terms of the elementary functions that are required to achieve its overall function or purpose" [93]. During this process, the overall function of the system is decomposed by smaller sub-functions in order to transform the system main function to alternative sub-functions that can be easily addressed by designers [102]. This modeling process explains the system architecture, structure and behavior. The generation of functional models during design processes provide designers with many benefits such as explicit response to customer needs, comprehensive understanding of the design problem, enhanced creativity and innovation [85].

Though originally developed for engineering systems, functional analysis is offered by Nagel et al. [85] as an abstraction tool for biomimetic design processes. Functional models of biological systems can be understood by designers who lack training in biological sciences. Nagel et al. analyzed several cases of biomimetic designs by functional modeling of both the biological and engineering systems and searched for the similarity and the differences between these functional models. Their research showed the applicability of the functional analysis approach for biomimetic design.

Vakili et al. [103] examined the role of functional models as stimuli for bio-inspired design. Participants were exposed to different biological representations, including functional models and visual representations. Results suggested that designers require an explicit list of all possible biological strategies, and these could be provided by functional models. Vakili et al. [103] suggested that functional models could represent biological strategies and save designers the efforts of extracting strategies. However, they reported that the creation of such functional models is difficult and therefore suggested to provide them to designers together with other stimuli for bio-inspired design.

Yen et al. [20] found that decomposing a particular function of a system into sub-functions improves the understanding of the interactions between the functions. Functional decomposition of the design problem and related biological system support the identification of the appropriate level for the analogical transfer [104].

Functional models may be presented by a graphic form of chains or trees. If all sub-functions are required to solve the overall function, then a chain form is required [93]. If each sub-function separately solves the overall function, then a tree form is adequate.

A function-means tree in engineering design spans the design space of possible design paths. Based on Hubka's law [105] of casual relations between functions and means, a function-means tree is a hierarchical structure of functions and means that realize the functions. The function at the higher level is the overall function (design challenge). The tree shows alternative solutions (means) to achieve the design challenge. Each mean, in turn, calls for further sub-functions to be realized by other means.

A function-means tree is used to analyze the design space, present design requirements and choose a preferable design path. A function-means tree was not proposed so far for biomimetic design as far as the authors know. A reason for that may be the interdisciplinary biomimetic context that requires moving backwards and forwards from biology to technology.

3.2.1.2 The C-K Modeling for Bio-inspiration

Salgueiredo [106] modelled the bio-inspiration process based on the framework of the C-K design theory [107]. According to this theory, design involves an interplay between the space of concepts (C) and the space of knowledge (K). Salgueiredo [106] referred to biology as well as to traditional engineering as the knowledge space and to technology as the concepts space, and explained the main roles of biological knowledge in the bio-inspired design process. Biological knowledge may (1) indicate a "design direction" (an expansion on the concepts space according to C-K theory); (2) indicate knowledge domains where no or limited knowledge is available; and (3) activate knowledge bases that would not otherwise be activated. This model can mainly direct designers to organize the knowledge explored in biology in a way that could lead the design process but it does not provide the knowledge itself and therefore can benefit from existing biomimetic databases.

3.2.1.3 List of Questions to Identify Suitable Analogies

Vakili and Shu [108] offer a list of questions that supports the understanding of the system function. For example: What does the system do? How is the system unique? How does the system carry out its function? These questions can guide the biological system exploration process but they do not provide a complete functional model of the biological system, as we aim to do.

3.2.1.4 The 'Causal Template' and 'Instructional Mapping Rules'

The BIDLab [109] developed the 'causal template' and 'instructional mapping rules' to increase the accuracy and quality of the analogy [110] based on a multi-step process:

- What is the desired function associated with the problem? What is the corresponding biological function?
- Which object in biology is acted on by the desired function?
- What is the precedent function that allows the desired function?
- In biology, identify the subject and object of the precedent function that you have identified above.

3.2.2 *Abstraction Methods Accompanied with Database Tools*

3.2.2.1 SBF Modeling and the DANE Interactive Computational Tool

SBF modeling [111] is an example of a functional modeling method that is based on the assumption that a complete representation of a system should include structural, functional and behavioral aspects. Structure is represented by components of a system and their connections. Function is represented as a schema of "before" and "after" implying the outcome of the system. Behavior is represented as a sequence of states and transitions between them that explain how the structure of the system accomplishes the function. The behavior actually defines the working mechanism of the system.

Yen et al. [20] suggested a simplified SBF approach and offered the WWH abstraction method based on What-Why-How questions. "What" identify the "Structure," "Why" identify the "Function," and "How" identify the "Behavior." Yen et al. [100] reported that the WWH model encouraged BID students to explore functional explanations of complex systems and extended their understanding of these systems. In addition, it facilitated interdisciplinary communication between biologists and engineering students.

However, framing the causal description of the system mechanism to fixed divisions of knowledge, function, structure and behavior, may lose the richness and complexity of the mechanism [112]. SBF or WWH abstraction is general and may not locate some aspects of the system functionality such as the source of energy or physical effects being used. In response, few attempts to expand the SBF representation model were suggested,

One extension is a representation framework called Structured Representation for Biologically Inspired Design (SR.BID) or the "Four Box Method", offered later by Yen et al. [20]. It expands the SBF representation and includes two more concepts: 'environment' and 'performance'. Both abstraction methods, the WWH and SR.BID, are meant to enhance understanding of the bio-inspired analogy. Generally, Yen et al. [20] reported that these representation methods helped students focus on the function of interest without being distracted by the other details.

Another extension of the SBF representation framework is the SBF Plus model, including two more basic concepts essential for representing system functionality: 'Source of energy, and 'Control'. Every functional mechanism requires a source of energy to operate and control system to stop or initiate the action. This model was offered by us (the authors) and it is based on a TRIZ system law, the law of system completeness. Further elaboration on this law may be found in Sect. 6.1.3.2.

3.2.2.1.1 DANE—Design by Analogy to Nature Engine [73]

DANE is an interactive computational tool developed by Goel [73] to facilitate biologically inspired design. DANE provides designers with the ability to create and maintain structured descriptions of biological systems. The structured representations is based on the Structure-Behavior-Function (SBF) ontology [111] to facilitate deep learning of biological system working principles and assist in the transfer of those principles into the design domain. It includes several biological mechanisms that are already analyzed by SBF model. DANE is also a research tool for better understanding analogical processes. The DANE tool was used by senior interdisciplinary class on biologically inspired design. This evaluation indicated its usefulness in supporting conceptual design of complex systems [20]. DANE supports the biomimetic design process by fostering interdisciplinary communication and by providing repository of designs for conceptually analogical system [113]. The tool is accessible via the net [73].

3.2.2.2 The SAPPhIRE Model and "Idea-Inspire" Software Tool

The SAPPhIRE model (State-Action-Part-Phenomenon-Input-oRgan-Effect) [114] explains the causal mechanisms for achieving a system function and is based on a function-behavior-structure approach. Function is defined as the intended effect of the system and behavior as the link between function and structure [112]. The model constructs are presented in Fig. 3.1. It supports an analysis of a system behavior based on a given structure, and it is suggested as a common, behavioral language for representing the functionality of both natural and artificial systems for supporting biomimetic innovations [112]. Sartori et al. [71] used the SAPPhIRE model to analyze 20 biomimetic examples and reported about its usefulness in representing complex biological systems and their functionality.

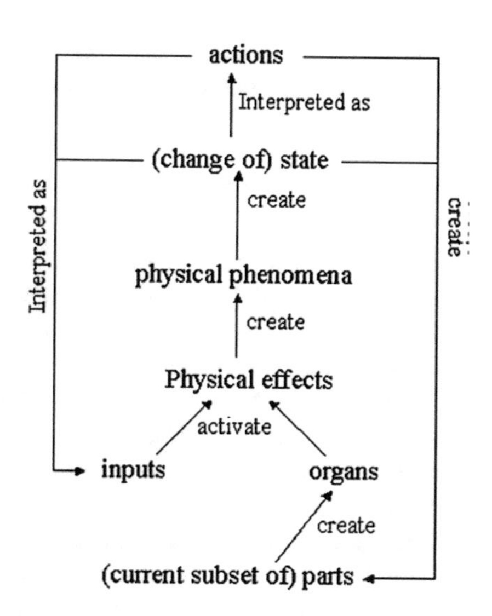

Fig. 3.1 The SAPPhIRE model of causality. Reproduced from [112] by permission

3.2.2.2.1 "Idea-Inspire" Software and Database Tool

Chakrabarti et al. [112] developed the "Idea-Inspire" software tool that allows designers to search a designated biomimetic database. This database includes about 100 natural and artificial systems represented by the SAPPhIRE model. The aim of the software is to inspire ideas mainly based on the behavioral aspects of natural phenomena. Designers can directly define the problem using the SAPPhIRE model constructs. They can also use software reasoning procedures for automated solutions search. The search is supported by a graphical user interface.

Chakrabarti et al. [112] evaluated the usefulness of the software for inspiring creative solutions by a preliminary experiment. They requested three designers with undergraduate degree in engineering to solve individually two problems with and without the "Idea-Inspire" software. They found that by using the software, all three designers got more ideas for solving each problem. While not statistically conclusive, it may indicate the potential of the software to foster creative solutions. The software is not accessible via the net.

3.2.2.3 The Contradictions Approach and BioTRIZ Database

Contradictions are part of reality and occur when design requirements contradict each other. Simply consider the requirements to improve functionality and reduce cost that are part of any project. Based on TRIZ, a physical contradiction exists if some aspect of a system must have two opposing states. A technical contradiction exists when improving one aspect of a design comes on the expense of another aspect. Viewing design problems as contradictions is a powerful way to observe problem from a new perspective [115]. Part of the TRIZ knowledge base is the contradiction matrix that summarizes how contradictions can be solved in technology.

3.2.2.3.1 The BioTRIZ database

Vincent et al. [24] used the contradiction approach to represent biological design conflicts and identify design principles that solve them. About 270 functions and 2500 conflicts were analyzed in an effort to identify biological solutions to engineering conflicts. The results are identified as inventive principles that solve those conflicts and summarized in the BioTRIZ database. Contrary to the basic contradiction TRIZ matrix [116], which contains 39 engineering parameters, the new BioTRIZ matrix is condensed and includes only 6 categories organized by the following principle: Things (substance, structure) do things (requiring energy and information) somewhere (space, time).

The contradiction approach may be used to represent biological design challenges. A designer who searches the BioTRIZ database can find sources of biological systems categorized by their inventive principles. The BioTRIZ database is not accessible via the net.

3.3 Transfer Methods

The last stage of the biomimetic design process is the transfer to technology. There are almost no dedicated methods to address especially this stage. There are definitions of what is being transferred (see Sect. 2.2.4) but almost no dedicated studies on this stage.

It may be assumed that the abstraction stage supports the transfer, as it precedes the transfer and defines clearly what should be transferred. When a designer understands what should be transferred, the transfer stage is done better. Although it is true, we still miss knowledge and tools to support this stage.

One way to support the transfer stage is to compare the biomimetic model with the biological model. This comparison can locate the similarity and the differences between the models and clarify what is similar and should be transferred and what is different and should be substituted. Nagel et al. [85] presented comparisons of biological and functional models, even though they did not present these comparisons as a tool to support the transfer stage.

Chapter 4
Literature Review Conclusions and Definition of Research Target

While in formation, the biomimetic discipline still lacks a clear theoretical and applicative framework of reference that could address the main challenge of the biomimetic process: bridging the gap between biology and technology. Efforts to address this challenge are described in the previous chapter. However, some gaps remain. The biomimetic design stages that are described in Chap. 2 are clear. The target now is to provide biomimetic designers with tools to follow these stages successfully.

4.1 Research Gap

The literature review reveals the following gaps.

4.1.1 Structure-Function Relations

The idea that function fits form or structure is one of the basic design principles in nature and well accepted both in biology and in the design literature. This idea is expressed by the idiom "*Function follows form/structure*" coined by Sullivan, Louis H. in 1896. Generally we may say that functions alone are more prominent in the previous efforts described in Chap. 3. Functions were identified as the common chain that connects biology and technology, an idea expressed by the idiom the "function bridge" [5]. The major searching approaches that are described in Sect. 3.1 are based on functional ontologies as meaningful searching keywords. However, the quality of the results is not ensured, and the search may end up with irrelevant analogies, as described in Table 3.2. Hill [99] indeed mentioned that associations of structures to principles may assist the searching process, but the structures or principles are not described in details. Searching biological databases by structures or

Y.H. Cohen and Y. Reich, *Biomimetic Design Method for Innovation and Sustainability*, DOI 10.1007/978-3-319-33997-9_4

structure-function keywords may yield more relevant analogies comparing to searching only by functions, particularly if we add details to the structure description.

Some of the abstraction methods that are described in Sect. 3.2 did acknowledge the importance of structure-function relations to biomimetic design. Sartori et al. [71] for example, referred to the need to identify the structure of the biological example that is related to the solution principles. The SBF modeling approach itself is based on these structure-function relations including the behavioural dimension. However, so far there are no tools to support these structure-function relations, i.e., tools to identify **specific** structures and **related** functions.

Compared to functions, structures are less subjected to personal interpretations and to domain terminologies. While biological and technological functions are derived from different terminologies, structures are visual and therefore less subordinated to different interpretations.

Structure-function relations may also provide a fertile platform for cooperation between engineers and biologists. This relation is inherent both in the biological and the engineering thinking and can be a common denominator between the two disciplines.

4.1.2 Patterns

Patterns are reusable solutions to recurring problems and accepted as a way to facilitate knowledge abstraction. They build analogies between observed solutions and problems yet to be solved and are used to transfer knowledge across domains [41]. Despite their potential to abstract knowledge and transfer knowledge between domains, they have barely been used for biomimetic design. The BioTRIZ [24] study is based on a patterns approach, but these patterns are limited to recurrent solutions for design contradictions.

The Life principles are actually based on nature sustainability patterns. The "Patterns from nature" project [41] is based on patterns in nature derived from ecosystems study. However, the results of this study are premature to be a foundation for a biomimetic design method.

The potential of a patterns approach has not been exhausted yet. Patterns may be the common chain that connects biology and technology, the "pattern bridge". Structure-function patterns in particular may abstract nature design solutions to various problems.

4.1.3 System View

Current abstraction methods for modeling and representation of biological systems are based on a partial system view. Functional modeling and SBF modeling do miss some system aspects such as the working structures, target object and energy source. (Some of these aspects may be identified during the behaviour analysis of the system but the model does not guide directly to identify them). The SAPPhIRE model [114] is based on a system view but it might be too complicated for some designers. The creation of biological functional models is reported as difficult [103]. There is a need for a simple and intuitive way to model biological systems.

Systems are inherent both in biology and in engineering and may be the common chain that connects biology and technology, the "system bridge". Knowledge base about systems such as general system theory [117] and TRIZ may infuse relevant knowledge required for the biomimetic design process.

4.1.4 Physical Effects

Physical laws and effects are principles of nature that govern a change [112]. They are sources of knowledge that can promote innovation of technological systems if applied during the conceptual design stage [118]. The application of physical effects for generating inventive solutions is the core of TRIZ knowledge base [115]. Physical laws and effects are essential factors of a system functional model that was mentioned in Sect. 4.1.3. So far, these laws and effects are barely integrated with current biomimetic design methods. The only abstraction method that referred to these effects is the SAPPhIRE model [114] that refers literally to the physical effect that creates the physical phenomena.

4.1.5 TRIZ Knowledge Base is not Exhausted

The TRIZ knowledge base is huge and has much to offer in addition to the contradiction matrix used as the base of the BioTRIZ study [24]. The TRIZ knowledge base has already been identified as a suitable core of knowledge to transfer knowledge between biology and engineering [119]. So far, not many efforts have been made to exhaust the potential of the TRIZ knowledge base for biomimetic purposes. A special core of TRIZ knowledge that is relevant to our discussion is the knowledge about technological systems, laws and patterns, that may be used to extend the system view of biological systems as a base for analogical reasoning, as suggested in Sect. 4.1.3.

4.1.6 Design Space Analysis by Functions and Means

Though functional modeling was offered by Nagel et al. [85] as an abstraction tool for biomimetic design, function trees were not studied widely for biomimetic design space analysis. Function trees in general, and function-means trees in particular may be useful for analyzing possible biomimetic design paths.

4.1.7 Sustainability

Sustainability is not integrated yet as an integral part of the biomimetic design process. Nature sustainability strategies, such as life principles, can address sustainability during early design stages. However, it is not clear how they were found and their application in engineering is not always straightforward. The application of sustainability tools in engineering might be clearer, or at least more straightforward, if presented through a technical lens, based on a technical system terminology and view.

There is a need to define a sustainability framework, based on a technical view, for searching applicable sustainability strategies in nature and to integrate it as an integral part of a biomimetic design method.

4.1.8 Transfer

The transfer stage of the biomimetic design process suffers from lack of study in relation to the other stages, the biological system searching and the abstraction stage. More studies on this stage could benefit biomimetic designers. Specific tools are missing at this stage.

4.1.9 Biomimicry as a Multidisciplinary and Interdisciplinary Design Process

The interdisciplinary and multidisciplinary aspects of the biomimetic design process were not discussed widely in the literature. The exact perspective each player (a biologist, an engineer, a designer, a marketing person and so on…) brings to the design process is related to the multidisciplinary aspect. The function each player has in the design process and how he or she integrates knowledge with others, is related to the interdisciplinary aspect. Both aspects should be studied and modelled to facilitate the biomimetic design process.

4.1.10 Biomimetic Problem Definition

The problem definition stage of the biomimetic design process could be studied more extensively. How should the problem be defined in a way that facilitates the core stages of the biomimetic design process? In what way is this biomimetic problem definition different from a non-biomimetic problem definition? There might be some problem definitions approaches and templates that have not been studied so far in relation to biomimetic problem definition.

4.2 Research Target

Based on the identified research gaps that are described in Sect. 4.1 we define the following research target to address parts of the identified gaps.

Develop a biomimetic design method for innovation and sustainability based on:

- Structure—Function patterns
- TRIZ knowledge base about systems and physical effects
- Nature sustainability patterns (integrated as part of the method)

The developed method is called: "The Structural Biomimetic Design Method". The research target responds to research gaps 4.11–4.18. We also aim to go beyond the level of statements and ideas and provide also proofs for the utility of the structural biomimetic design method. We do it by experimentations and case studies performed not only in lab but also in the field, addressing a real design challenge.

Part II
Research Method

Chapter 5
Research Model

In this chapter we present the research model that we followed during the development of the structural biomimetic design method.

5.1 Definitions

The design literature is rich in various concepts. The following definitions describe some of the major concepts that construct this domain. Their hierarchy and their relations are presented in Fig. 5.1. We avoided more detailed descriptions of these relations for the sake of simplicity. Guidelines for example, appear as part of a design method but may also direct which design method to use in a given case.

5.1.1 Design Methodology

Design methodolgy, is mainly related to "The study of how designers work and think, the establishment of appropriate structures for the design process, the development and application of new design methods, techniques and procedures, and reflection on the nature and extent of design knowledge and its application to design problems" [120]. But such study always lives in a particular worldview which embeds values and goals of the holders of the worldview [121]. Hence, there could be differing design methodologies depending on their worldviews or even different methodologies within a single worldview. Design methodologieson one hand support the development of design processes, methods and tools and on the other hand they are enriched and refined by these development activities.

Y.H. Cohen and Y. Reich, *Biomimetic Design Method for Innovation and Sustainability*, DOI 10.1007/978-3-319-33997-9_5

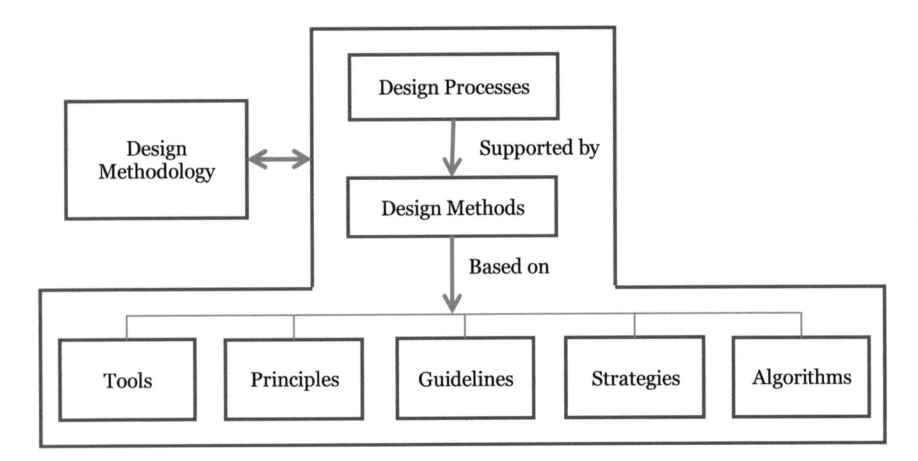

Fig. 5.1 Hierarchy of, and relation between, major definitions in design

5.1.2 Design Process

Design is a process of developing a new object, a solution, to achieve a desired goal. It could be described at different levels of abstraction including high level steps that should be followed in order to achieve the design target or a record of the detailed actions that were involved in a particular process. Design processes are supported by design methods. Each design stage with its multidisciplinary basis might require different design methods as demonstrated by the study of Kolberg et al. [122].

5.1.3 Design Method

A design method is a technique to direct design processes based on algorithms, tools, principles, guidelines, and strategies that assist designers to achieve the design target.

5.1.4 Design Tools

Design tools help designers to accomplish actions or tasks. They may help in representing knowledge, enriching information, generating new solutions and decision making. Some design tools are developed especially for a specific design method while others already exist and may support various design methods. For a particular task, such as selection between alternatives, there is no necessarily single design tool better than another, and some tasks may involve more than one design tool [123].

According to these definitions, we aim to develop a biomimetic design method that supports the biomimetic design process. For this purpose we might need to develop design tools.

5.2 Developing a Design Method

Design methods are required to enhance the development of new products, systems and technologies. These methods guide the designer intuition, support decision making in various stages and finally meet the design challenges. Several characteristics of design methods are desirable including: (i) Validity—achieving the declared objectives; (ii) Simplicity—based on a limited number of patterns or operators; and (iii) Easy to learn—applicable after a short training.

Development of a design method is first of all a design activity, where the object of designing is a method and not a product. It is a complex activity that starts with formulating the problem, identifying requirements, searching for solutions and finally, evaluation and validation of these solutions. This development process becomes more complicated when the design methods are developed in an interdisciplinary and multidisciplinary environment, such as biomimetic design. Despite the variety of design methods and tools that have been developed so far, there is lack of systematic ways to support the process of design method development. While a regular design process of a new product may lean on existing knowledge bases, design process of a new design method may require the construction of new knowledge [124].

5.3 Research Model

For the purpose of knowledge construction, we followed the model described in Fig. 5.2. There are several approaches to research models for the purpose of developing new methods or systems. One model was offered by Nunamaker et al. [125] for information systems development. Another model was offered by Reich et al. [126] for the purpose of evolving theories, tools and solutions for design. These models reflect our design philosophy of iterative study and development; design that is based on an interplay between current knowledge source of observations, repositories, theories and methods, with the process of project (system/method) development, while the evaluation of the project is used in turn, to refine the current source of knowledge. Such interplay between the newly developed designed method and current and newly developed knowledge bases is the core of our research model, presented in Fig. 5.2. The uniqueness of this model is its relation to the development of design methods and not only systems or projects,

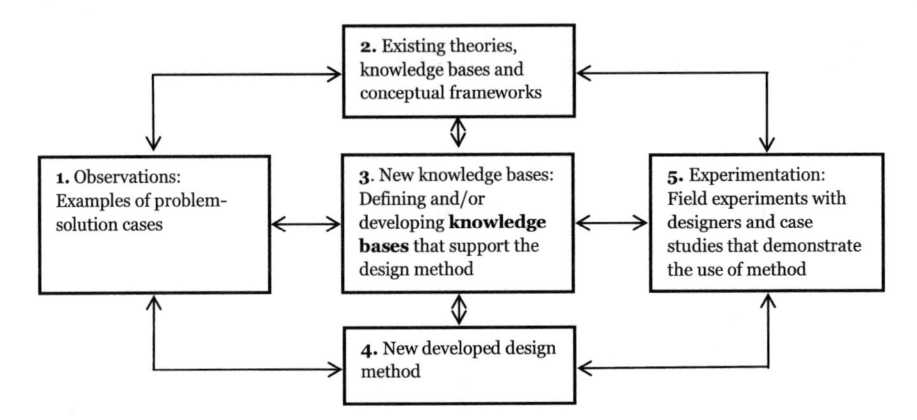

Fig. 5.2 Research model: knowledge construction for a new design method

but moreover, to the development of new knowledge bases that support the new methods. Note that not all newly developed design methods require the construction of new knowledge bases. In other cases, a new knowledge base sets an opportunity for the development of a new design method [124].

This model led us towards the development of the three bases of the structural biomimetic design method, presented in Chaps. 7–9. These bases were later integrated within the design process as presented in the structural biomimetic design method manual in Chap. 10. The structural biomimetic design manual was later validated by case studies and field experiments, as presented in Chaps. 11 and 12.

5.3.1 Explanation of Research Modules

The purpose of this model is to lead the process of knowledge construction as a foundation of a new design method. Such new knowledge could be translated into tools, guidelines, strategies, principles and algorithms of a new design method that is evaluated by experimentations. According to our research model (Fig. 5.11), establishing a new design method involves integrated top down and bottom up processes. On one hand we gain bottom up insights from observations of problem-solutions examples. These insights are understood in light of the knowledge derived from existing theories and knowledge bases. On the other hand, existing knowledge and theories can guide us where to search for observations of problem-solution cases. The new knowledge bases are derived from our insights regarding the observations and the current existing knowledge, but once it is defined, it could enrich them back. The new knowledge bases affect also the experimentations plan and are also refined by their results. This interplay between the model constructs is presented by the bidirectional arrows in the model.

5.3.1.1 Observations (Module 1)

Observing cases of problems and solutions can provide the essence of the problems domain and type of solutions. For this purpose, there is a need to identify the entity that bears the problem. These observations can give clues on related knowledge (module 2 and 3) and even give a glance at the possible design method (module 4).

5.3.1.2 Existing Theories, Knowledge Bases and Conceptual Frameworks (Module 2)

The newly developed design method is related to some disciplines. There might be theories, knowledge bases and conceptual frameworks that explain phenomena and processes we aim to address by this newly developed design method. Identifying this core of knowledge is a result of a literature review. While the general literature review presented in part I located previous efforts and defined the gap that is needed to be addressed, here we aim to identify knowledge that provide understanding regarding the problem-solution cases (module 1); define the knowledge bases that support the design method (module 3); and explain the experimentation results (module 5).

5.3.1.3 Knowledge Bases of the New Design Method (Module 3)

The interplay of literature, observations and experimentations yields a solid foundation of knowledge that can support the new design method (module 4). It may exist already or should be newly elaborated for the developed design method. This knowledge is located by a literature review (module 2) and the problem-solution observations (module 1) while it is sharpened and refined by the experimentation (module 5). On the other hand, it enriches our understanding of modules 1, 2 and 4.

5.3.1.4 Newly Developed Design Method (Module 4)

The knowledge bases of the newly design method (module 3) is integrated within a design algorithm formulated by a manual. The design method manual is examined by several case studies and design experiments that refine both the design method (module 4) and the structure of knowledge itself (module 3). Thus, establishing a design method is an iterative process of interplay between theories, observations, and experimentations.

5.3.1.5 Experimentation (Module 5)

Experimentations serve both the purpose of refining the knowledge bases by getting feedback from their use and the purpose of validation, reassuring that the design method achieved its declared objectives. Experimentations include case studies and field experiments.

- Case study as a research method

Case study as a research method is common in a wide variety of fields, including design research. It is used for theory building [127] and as a research strategy [128]. Case studies are suggested when a researcher aim to answer 'why' and 'how' questions, mainly for complex issues. Then, case studies provide detailed contextual analysis of a limited number of events referring to real-life context. Limitations of case studies are mainly their small number that might limit the generality of relevant conclusions [128].

In our research, case studies are examples of detailed solutions to real challenges achieved by the structural biomimetic design method. Case studies provide a go/no go validation for the design method as they demonstrate a transfer of knowledge from biology to technology. In addition, case studies refine of the design stages as they identify the critical stages, reveal difficulties, and identify needs for additional design tools to support the design process. Selected case studies are presented in Chap. 11.

- Field and lab experiments for design method validation

Field and lab experiments provide a wider view of the design method performance. While case studies are limited to small number of cases, experiments are based on a larger database. Whereas lab experiments have a quantity advantage, the solutions are less developed and enriched. However, conclusions may go beyond the go/no go stage, as the solutions may be assessed by several criteria and compared statistically to other design methods. The field and lab experiments are presented in Chap. 12.

5.4 Implementation of the Research Model for Biomimetic Design

We followed the research model (Fig. 5.2) and applied it for developing a biomimetic design method. The application of the research model for biomimetic design is presented in Fig. 5.3.

Module 1: Observations

Our research model (Fig. 5.3) guides us to locate pairs of biomimetic applications and biological systems. These pairs capture the edges of the biomimetic design process, the biological source and the target application. Our database was constructed in two stages. First, we collected examples of biomimetic applications extracted from different biomimetic sources [5, 87, 129]. For each biomimetic application we located the biological organism that served as the source of knowledge for inspiration or imitation.

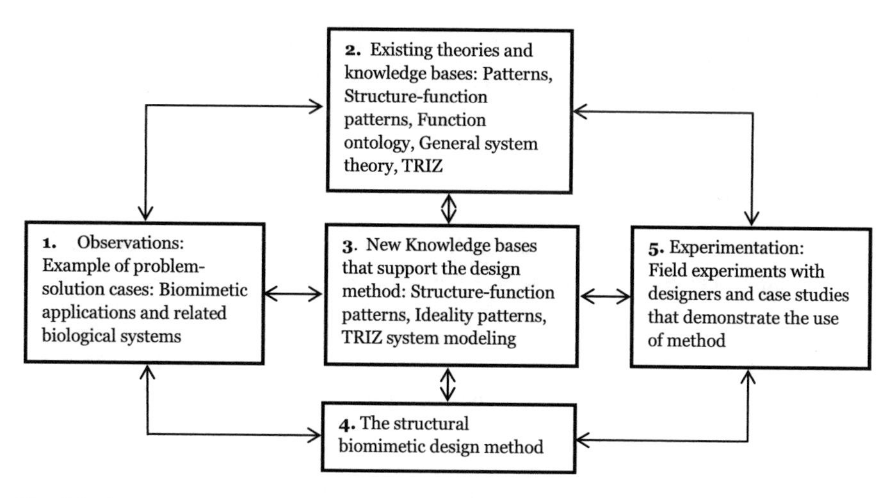

Fig. 5.3 Application of the research model for biomimetic design

In the second stage, we added many more biological systems, not necessarily related to biomimetic designs to verify our findings in a broader scope of general biological systems. In total, our database includes 140 biological systems.

Following our research model we focused on the problem-solution relations. For each biological system we studied the design problem demonstrated and the biological solution that was later transferred to the biomimetic application. For this purpose we analyzed the biological systems in our database with tools identified in module 2 of our research model.

Module 2: Theories and Knowledge Bases

The analysis of the biological systems in the observations database was based on several theories and knowledge bases, including general system theory, TRIZ, patterns approach, structure-function patterns, and functions ontologies. The location of these theories and knowledge bases is a result of the identified research gaps presented in Chap. 4. Elaboration on this module is presented in Chap. 6.

Module 3: Knowledge Bases that Support the Design Method

The analysis of the biological systems in our observations database (research database) provides the three bases of the structural biomimetic design method.

Base 1 Functional Patterns
Base 2 Structure-Function Patterns
Base 3 Sustainability patterns

Module 3, the knowledge bases that support the structural biomimetic design method, is presented in Chap. 7–9.

Module 4: The Structural Biomimetic Design Method

The integration of these knowledge bases into a coherent design method is summarized by the structural biomimetic design method manual and presented in Chap. 10, including the design algorithm and the tools.

Module 5: Experimentation

The experimentation and validation of the design method by case studies and field and lab experiments are presented in Chaps. 11 and 12.

Chapter 6
Theories, Knowledge Bases and Conceptual Frameworks that Support the Analysis of Observations

In this chapter we present the theories, knowledge bases and conceptual frameworks (module 2 of the research model) that support the analysis of our observations (module 1 of the research model).

6.1 The Technical Lens Approach for Analyzing Biological Systems

In part I of this book, we elaborated the gap between biology and technology that is derived from different terminology and way of thinking (Sect. 2.6). Therefore, it is clear that an engineer or designer who has no previous knowledge in biology encounters difficulties to analyse and understand biological systems. Now, let's assume we could observe biological systems as if they were technological systems by using technical knowledge about systems. Such observation is innovative and may provide designers with a modeling and abstraction tool to analyze biological systems. This technical approach for observing biological systems is supported by the following theories and knowledge bases.

6.1.1 Systems

6.1.1.1 General System Theory

General systems theory was proposed in 1936 by the biologist Ludwig von Bertalanffy [117]. He suggested that the same system concepts, organizing principles models and laws are common in different disciplines including physics, biology, technology and even sociology. Bertalanffy defined a system as a complex of elements in interaction. Later definitions of systems added the idea of function stating that a system is a

Y.H. Cohen and Y. Reich, *Biomimetic Design Method for Innovation and Sustainability*, DOI 10.1007/978-3-319-33997-9_6

collection of parts which interact with each other to function as a whole [130], or a set of things that produce their own pattern or behavior over time [131].

Bertalanffy offered the principles that are valid for 'systems' in general and not depend on the type of system, its elements or the relation between them. It is interesting that in fact a biologist realized a general framework to understand systems, their relation to the environment and the relations between their parts.

Biological systems have been evolving for millions of years while technological systems have been evolving for only a few hundred years. There are major differences between biological and technological systems. Technological systems are designed and evolved by humans for performing functions. In contrast, biological systems evolve with their genetic codes governed by natural selection. Notwithstanding the major differences between biological and technological systems, they are both systems, and as such, they share some common features, as Bertalanffy suggested [117].

Adopting a system view during the biomimetic design process provides a platform for analogical transfer from biology to technology. The resemblance of biological systems to technological systems sets a platform for identifying analogies between domains. This innovative view may provide new knowledge about the qualities of biological systems, where this knowledge is presented in a technical language that is familiar to engineers.

6.1.1.2 System Hierarchy, Parts and Boundaries

A TRIZ [132] view suggests observing systems at several levels:

- System level: the system itself.
- Supersystem level: Environmental components that do not belong to the system but interact with the system components and have an influence on system functions.
- Subsystem level: Components that are parts of the system assembly and are essential for its function.

Adopting this broader view of system hierarchy and boundaries may provide a clear platform for system analogical transfer from biology to technology. The process analysis according to this hierarchy, including definitions, guiding questions and examples, is demonstrated in Table 10.6, presented in Sect. 10.2.4.

6.1.2 Functions

6.1.2.1 Functions in Technology and Biology

Function is a basic characteristic of an entity. In relation to technological systems, functions refer to their goal or purpose. Technological systems are designed by humans for performing functions. However, a goal cannot be associated to biological systems as they are evolved by the laws of nature.

A biological function is a function that is: (i) part of an organism and (ii) has a physical structure as a result of the coordinated expression of that organism's structural genes [133]. Each function has a bearer with a physical structure. In the biological case, the bearer has naturally evolved to have, or, in the artifact case, the bearer has been designed to have [134].

In technological systems, function is referred to as the goal of the design. In biological systems, the functional mechanism was not predetermined or planned in advance and can be sometimes a random genetic evolutionary change. However, in biological systems we can identify what they actually do. Therefore, when we use the word function in relation to biological systems we actually should refer to an interpretation of the system actual behavior or to the work that it does.

6.1.2.2 Functional Modeling of Systems

Functions are key elements in designing systems. Functional modeling of systems explains the system architecture, structure and behavior [135]. Some approaches suggest analyzing the flows of the system as well [85]. Elaboration on functional modeling appears in Sect. 3.2.1.1.

Functional modeling of biological systems has already been introduces as a framework for analyzing biological systems as a base for analogical transfer to biomimetic application [85, 93]. The engineering-to-biology (E2B) [95] thesaurus that relates biological function and flow to engineering terms could enhance this process. The generation of functional models of biological systems is expected to enhance innovation and provide designers with better understanding of the design problem in general and of the functional mechanisms in particular [85].

6.1.2.3 Functions Ontologies

In order to support functional modeling processes, there is a need for functional ontology. Different representations of the same functions will yield different ways to represent the same design concept. This need for formalized function ontology is prominent especially during interdisciplinary design processes, when the inconsistency of functional vocabulary across domains may lead to ambiguity, complexity, non-uniformity and non-repeatability of functional models [136–138]. Inconsistencies even exist in one discipline such as mechanical engineering, so involving additional disciplines exacerbates the situation. Lack of domain independent functional ontological framework that provides the most general domain-independent functions led to developments of top level function ontologies that describe the most general functions present across domains [139]. These ontologies aim to identify a minimal set of functions that do not overlap and yet span the space of designed requirement [137–139].

Latest effort to standardize these ontologies by the NIST design repository project [137] still lacks a confirmed relation to biological functions. A top level function ontology that is derived from biological systems and encompass biological functions could support biomimetic design processes.

6.1.3 TRIZ—Inventive Problem Solving Theory

TRIZ is a "theory of inventive problem-solving" derived from the study of thinking patterns used to model inventions that are found in the global patent literature [23]. TRIZ knowledge base is derived from the study of hundreds of thousands of technological systems, and as such it contains extended knowledge about the structure and nature of technological systems. It mainly suggests 40 inventive principles to solve contradictions in technological systems, laws of technical system evolution, system modeling tools (Su-Field) and algorithm of inventive problems solving (ARIZ).

TRIZ tools were related to developing eco-design tools and frameworks for sustainable design. Russo et al. [140] used some TRIZ tools such as Ideality, Resources and Laws of system evolution, to form several practical eco-guidelines for product innovation and sustainability. They claimed that TRIZ tools benefit eco-design mainly by providing guidelines to increase product sustainability and a better usage of resources. In addition, TRIZ tools extend the way we evaluate resources taking into consideration their contribution to the system function and their potential effects with other supersystem factors. D'Anna and Cascini [141] offered the SUSTAINability map, a new approach towards the preliminary analysis of sustainability problems. It is based on two basic TRIZ tools: the Laws of system evolution and the System Operator.

In relation to biomimicry, TRIZ has already been identified as a main core of knowledge suitable for bridging the gap between biology and engineering [119]. This transfer process between domains benefits both our understanding of biological systems and increases TRIZ knowledge with new biological inventive principles. Altshuller himself identified the potential of exploring the 3.8 billion years of natural design lab to improve and enrich the TRIZ Method [23]. But Altshuller's vision was not fulfilled until the latest study of BioTriz [24], as most of the TRIZ knowledge until then was based on different fields of engineering including physics, chemistry, and mechanical engineering, with no reference to the biological base of knowledge.

One of the main modules of TRIZ knowledge base is the Laws of system evolution. Altshuller [23] studied the development path of technological systems over time and discovered few trends. He formulated these evolutionary trends in a group of laws regarding the being, operation, or change of technological systems. These laws are mainly used of forecasting trends in the system development but they also provide a good sense of the system present and future structure.

We identified one of these laws, the law of system completeness, as suitable for the purpose of modeling and representing the **structural aspect** of biological system. We identified another TRIZ modeling tool (Su-Field) as a suitable tool for modeling the **functional aspect** of a biological system. By modeling biological systems with these TRIZ tools we can provide a functional model of the biological system and explain the role of structures within a complete system that has a main function. Another law, the law of ideality, was identified as suitable for the purpose of extracting sustainability patterns from biological systems. As far as the authors know, an analysis of biological system by these TRIZ laws and tools has not been done yet, and therefore may be a source of innovation.

6.1.3.1 Su-Field Analysis Model

Substance-Field (Su-Field) is a TRIZ analytical tool for modeling problems [142]. The authors used this tool not for the purpose of modeling problems but for the purpose of defining the main function of the biological system. According to this model, shown in Fig. 6.1, a function is defined as an interaction between two substances and one field. The working unit (substance 1) acts on the target object (substance 2) via a field of energy to perform a certain function (action). We identified the two substances and the field of energy for each biological system in our observations database.

6.1.3.2 The "Law of System Completeness"

This law describes four essential elements of a technological system: engine, transmission unit, working unit and a control unit (Fig. 6.2). The engine converts

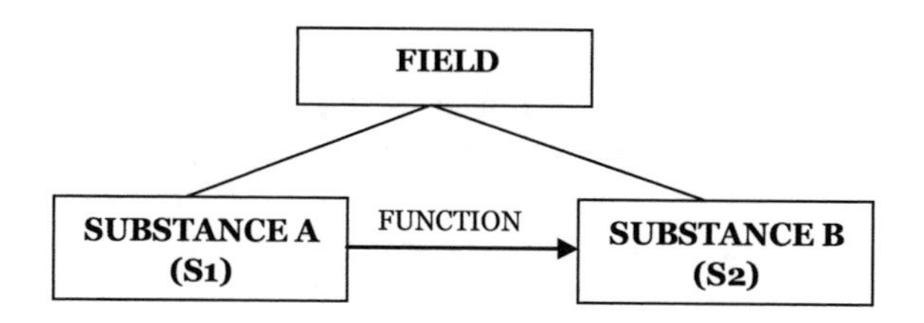

Fig. 6.1 Substance-field definition of function

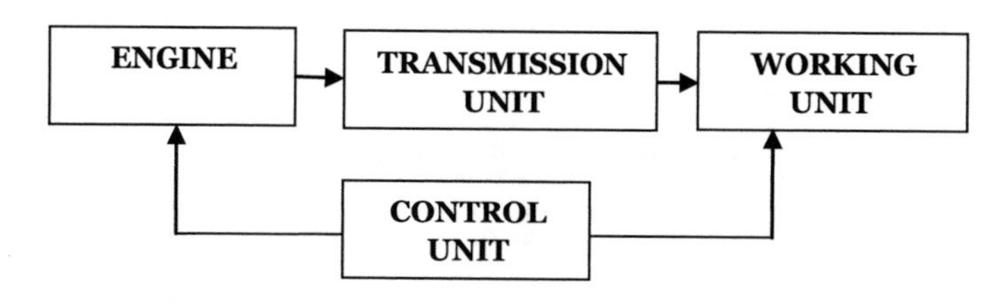

Fig. 6.2 The law of system completeness model

the energy, the transmission unit transmits the energy, the working unit performs the main function of the system, and finally the control unit guides and controls the parameters that are changed by the function. The law requires that all components are present and that if any component fails or is missing, the system does not survive [143].

We analyzed each one of the biological systems in our observations database using the law of system completeness identifying the main system elements, i.e., engine, transmission unit, working unit and a control unit in a the biological system. These elements are components of the biological subsystem or supersystem.

6.1.3.3 The Complete System Viable Model

The integration of the law of system completeness and the Su-Field functional definition provides a complete viable system model, presented in Fig. 6.3. It is clear that some aspects are missing at the law of system completeness model such as the system source of energy and the target object of the system function. These aspects are provided by the Su-Field model. The working unit is a common factor that appears in both models. It is the first substance of the Su-Field model that performs the action and the working unit of the law of system completeness.

Mann [144] introduced the integration of "the law of system completeness" and the "Su-Field" analysis as the complete viable system model, providing a broader view of the system structure in the context of its function. This model can provide us with a method addressing both the structural and the functional aspects of biological systems.

6.1.3.4 Ideality

Ideality is derived from the word "idea" implying a state of imaginary that exist only as an idea but should be endeavored. It is also implying the quality or state of being ideal, a standard of perfection, beauty, or excellence (Merriam–Webster dictionary). According to TRIZ, ideality is defined as the qualitative ratio of system

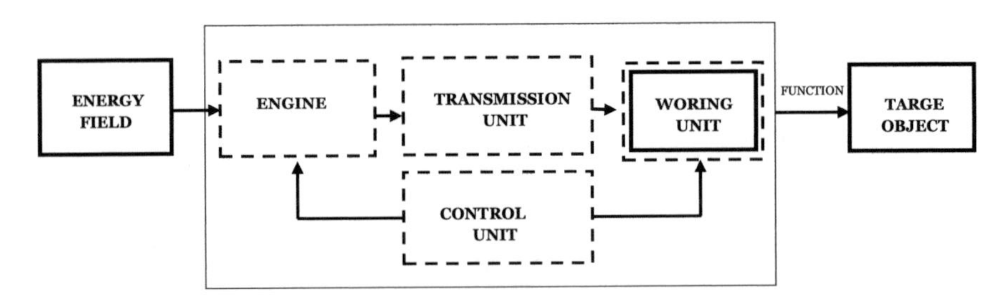

Fig. 6.3 The complete viable system model: *Bold line* elements are part of the Su-Field model. *Dashed line* elements are part of the law of system completeness

$$\text{Ideality} = \frac{\text{All } \textbf{Useful} \text{ functions}}{\text{All } \textbf{Harmful} \text{ functions}} = \frac{\text{Benefits}}{\text{Costs}} \rightarrow \infty$$

Fig. 6.4 The law of ideality

useful functions to its harmful functions [145], as presented in Fig. 6.4. Useful functions are the benefits that the system provides and harmful functions are the undesired costs of the system operation, such as resources, noise, waste, pollution, etc. Any technological system tends to become more ideal throughout its lifetime, providing more benefits in less costs. The ideal final result is a hypothetical state of having all benefits at zero cost. This hypothetical situation cannot be achieved but it directs the evolution of the system towards ideality [23].

Summary
The technical lens approach for analyzing biological systems is based on viewing biological systems as technological systems. This view follows general system theory [117] stating that the same system concepts, organizing principles, models and laws are common in different disciplines. The TRIZ knowledge base regarding technological systems may be used for analyzing biological systems under a technical lens, even though it was originally developed for technological systems.

One level of applying this technical lens is the functional level, observing the biological system functional mechanism as if it was a technological system. Functional analysis of biological system can be supported by functional ontologies or by technological functional models such as the TRIZ Su-Field model. Another level of applying this technical lens is the structural level, observing the components of biological system as if they were a part of a technological system. The last level of applying this technical lens is the sustainability level, observing the sustainability aspects of biological systems through the ideality framework. Analysis of biological systems under these technical lenses provides an operational language that eases the description of these systems and improves their understanding.

6.2 The Patterns Approach for Analyzing Biological Systems

In the previous section we introduced the technical lens approach for analyzing biological systems. While performing this analysis of biological systems, we searched for patterns. In this section we introduce the patterns approach for analyzing biological systems.

6.2.1 What Are Patterns?

Patterns are practice-proven, re-usable solutions to recurring problems [146]. The term ‘pattern’ as used by Alexander [147] refers to models or archetypes that can be followed in making things. He defined that “Each pattern is a three part rule which expresses a relation between a certain context, a problem and a solution” [148]. He claimed that design issues can be solved by integrating coherent and modular solutions to specific problems. Alexander developed the concept of ‘pattern language’ [148] in the late 1970s as a mean of capturing and communicating recurrent problems and solutions in architecture. He described a set of 253 patterns that capture the key aspects of living buildings. Patterns were adapted later in other fields beyond architecture such as software engineering [149] or project management [150].

Patterns are harvested from practice and therefore are observed and not invented or developed [146]. Most pattern languages use a ‘bottoms-up’ approach by collecting large numbers of problem/solution sets and analysing them to extract the patterns.

Pattern may be described by several characteristics [146, 151] such as: Re-usable solutions, target problem, known usages and more. Patterns are rarely used independently and their integration captures solutions for more complex problems.

6.2.2 Patterns Based Design Method

Patterns are accepted as a way to facilitate knowledge abstraction. In relation to design, patterns are an abstraction of design solutions that are relevant for a group of design problems [148]. They build analogies between observed solutions and problems yet to be solved and are used to transfer knowledge across domains [41].

Patterns may be a base for design method when they provide clear solutions to identified problems. In this case, patterns allow designers to disconnect from the details and use the abstracted problem-solution knowledge during design processes [44]. Each of the patterns may be implemented individually, with no dependence on the others.

Patterns do not provide us with the richness and complexity of the design space and do not guarantee finding a solution. However, they embed some design principles and expand the space of potential solutions that would be difficult to find otherwise. In some cases the pattern is not the full solution but a clue for a general solution direction that needs to be further investigated.

Experience proves that analysing repeated patterns in a group of solutions could lead to developing of a new design method. There are several examples of pattern based design methods. Altshuller [23] analyzed about hundreds of thousand patents in order to find repeated patterns that characterize systematic inventive thinking. This analysis led to identifying repeated solutions to engineering conflicts summarized in the TRIZ contradiction matrix. This method enables to detach the problem from its context and base the solution on repeated principles. Another example of developing a method by searching a large number of engineering inventions is the development of ASIT [152]. ASIT emerged by analyzing many solutions that satisfy the Closed World and Qualitative Change conditions. For each solution, Horowitz examined the main modifications made to engineering systems in the transition from the problem to the solution world. This analysis led to identifying five main creativity enhancing operators: unification, multiplication, division, breaking symmetry, and object removal.

Both Altshuller and Horowitz started from a general group of engineering solutions and located repeated patterns. Without finding these patterns, both of them might not have proceeded towards a consolidation of a method. The axioms of Axiomatic Design (AD) have also been described as arising from the analysis of good quality products [153].

These three examples are of course not a representative sample, but there is no chance to find a large sample within the methods domains. At the meta-level, we may assume that in order to proceed towards a biomimetic design method one should explore and analyse the group of biomimetic developments in order to find repeated patterns. As TRIZ, ASIT, or AD cannot guarantee solutions, but they facilitate focusing on the core problem issues and directing towards potential solutions, so will be the new method.

In relation to biomimetic design, Bhatta and Goel [154, 155] suggested that design patterns are one of the fundamental units of analogical transfer between the source (biology) and the target (technology) of the biomimetic cross domain analogies. Following this study, Goel et al. [44] presented how biologically inspired sustainable design can be analyzed in terms of cross-domain analogical transfer of design patterns.

6.2.3 Patterns and Biomimicry

The essence of biomimicry is the process of imitating nature design solutions. Patterns are "forms or models proposed for **imitation**" [Merriam Webster]. Alexander himself referred to patterns as models or archetypes that can be **followed** in making things, so it is reasonable to use patterns when aiming to imitate or follow nature solutions. Indeed, there are several examples for using nature patterns for the purpose of promoting biomimetic design processes including the Life principles (Sect. 1.6.3.1), the BioTriz study (Sect. 3.2.2.3) and the "Patterns from nature" project [41]; however, the potential of patterns to be a base of a biomimetic design method, has not been exhausted yet, as described in Sect. 4.1.2.

6.2.4 Structure-Function Patterns

Nature forms and structures provide a wide range of properties with minimal costs. Form is usually related to the shape of an object, while structure is considered as the arrangement of its parts. Both of them have been studied by researchers from various disciplines [156–162] that acknowledged the appearance of structural and form patterns in nature. These patterns are forms or structures in which some features recur, identically or similarly in places that apparently have nothing in common [156], and provide benefits and efficiencies, irrespective of their size or material [160].

The consented explanation of the origin of forms and structures in nature relates them to the forces in space. D'Arcy Thompson [161], a pioneer in the research of organic forms, suggested that nature forms are obedient to the laws of physics and are a product of the "invisible" forces in space. The physicist Richard P. Feynman, a Nobel Prize winner, suggested that natural objects respond directly to the lines of force acting in their common space [163]. A similar idea was introduced by Steven Vogel, a biologist who argued that all organisms are exposed to the rules of the physical sciences that constrain the range of designs available for living systems [164]. The idea that a form is a result of the physical forces has been used to develop the Constructal Theory of Bejan, who claimed that a geometric form (shape and structure) is a result of the struggle for better performance under global and local constraints [8]. As the range of possible designs is constrained by space forces and nature laws, we may see some repeated forms and structures as patterns. These patterns are the design solutions to problems that are set by the constraints of space.

However, structures and forms do not stand alone. They are related to functions. The idea that function fits form or structure is one of the basic design principles in nature and well accepted both in biology and in the design literature. A peculiar coherence between a living structure and a function it performs is referred as design among biologists [165]. The concept of function is subject to confusions and is difficult to define in biology, but it is generally defined as the uses or actions of structures [166]. Although this relation might be complex [167], at the meta-level, we might find general relations between structures and functions in nature.

As a result of this review, it is clear that we may look at nature forms and structures as solutions to functional requirements under space constraints. Thus, the pairs of structure-function represent cases of problem-solution, whereas the structure provides the answer for the problem: How can one achieve a specific functional quality under space constraints? For the sake of simplicity we use the term structure-function but it includes the term form-structure too.

Structure-function relations are interesting from biomimetic point of view. While morphology is the study of the shape and structure of parts of an object and how they create the whole [168], functional morphology is the study of their relation to function and is considered as one of the oldest strands of development in biomimetics [4]. Better understanding of the processes that affect morphology in biological systems was defined as the next phase of biomimetics [169]. A designer that considers only the choice of materials for specific functions misses the enormous effect that structures do have on functions [160]. Understanding nature structure-function patterns may lead to choices of efficient and sustainable structures.

Indeed, several approaches to analyze biological systems in the context of biomimetic design lean on these structure-function relations. Sartori et al. [71] mentioned the identification of function, solution principle and structure as guidelines for biological system analysis. Goel offered the Structure-Behavior-Function (SBF) approach [111], including also the behavioral aspect, to facilitate deep learning of biological systems and incorporated it in the DANE computational tool for design by analogy [73]. Chakrabarti et al. [112] offered the SAPPhIRE model—another Structure-Behavior-Function approach—to explain the causality of natural and technological systems.

One approach to investigate these structure-function relations from a biomimetic perspective is to conduct a thorough analysis of this relation. Fratzl [25] suggested that a thorough analysis of structure-function relations in natural tissues should be performed in the process of a new bio-inspired material. Gorb [170] provided a detailed review of the relations of biological surfaces and functions, though he mentioned that due to the complexity of biological surfaces, only few exact working mechanisms have been clarified so far. He elaborated on several functions such as antifriction and drag reduction, adhesion, filtering and self-cleaning and described several mechanisms to achieve each function. For example, he elaborated on mechanical and chemical attachment mechanisms including hooks, suction and secretions [171].

Another approach to investigate these structure-function relations from a biomimetic perspective is to search for patterns that abstract this relation. Whereas the thorough approach to investigating these structure-function relations provides the details and exact mechanisms behind them, we aimed to zoom away and extract some general structure-function patterns that provide an abstraction of nature structural solutions.

For this purpose we have to give up the richness and complexity of the different morphologies and related functions as described by Gorb [170] and strive for some generic patterns of structure-function relations. In doing so, we included our observations on general forms and structures in nature and not only on surfaces, as Gorb did.

The patterns approach spans the space of structure-function relations on two dimensions, as if it was a flat space, while the thorough approach with its detailed studies provides a third dimension of depth. Both approaches are required to complete a successful biomimetic design process. The patterns approach supports the abstraction stage of the biomimetic design process, when design principles are extracted based on understanding of the biological mechanism [54, 56], while the thorough approach provides the particular knowledge and details that are required for the implementation of the biomimetic application.

Summary

Based on the patterns approach that can provide us "forms or models proposed for imitation" (Merriam Webster definition of pattern), we searched for patterns when analyzing biological systems under the technical lens. Patterns support the abstraction stage of the biomimetic design process and were identified as one of the fundamental units of the biomimetic analogical transfer. We mainly focused on structure-function patterns as pairs of solution-problem cases, where the structure provides the solution for a design problem set by the constraints of space. Nature structure-function patterns may be a base of a patterns based biomimetic design method and serve as an index of clues for extended information on the richness and complexity of these structure-function patterns.

Part III
Research Methodology, Process and Results

In this part, we introduce the three knowledge bases of the structural biomimetic design method, according to module 3 of our research model (Fig. 5.3). These knowledge bases are a result of various analyses of biological systems derived from our observations database (module 1). The results of these analyses are structure–function patterns and ideality patterns of biological systems.

Research Methodology

The purpose of this part of the study may be formulated as a research hypothesis.

Hypothesis There exists an abstraction level in which biological systems could be described with a small set of structure–function patterns or ideality/sustainability patterns. A small set is better following the parsimony or simplicity principle in science; a small model or simple explanation is always better than a complex one, given that otherwise they are similar. The term "small" in the hypothesis should be interpreted as not larger than the size of existing functional bases developed in design research [121], or the size of sustainability bases, such as life principles [5].

This research hypothesis calls for finding an appropriate abstraction level and modeling tool and using them to describe biological systems with a small set of structure–function patterns or ideality patterns. In a traditional study, one would articulate a hypothesis and then create a representative sample to experiment and conduct some statistical test to reject a null hypothesis which in our case would be that "there is no abstraction level in which biological systems could be represented with a small set of structure-function patterns or ideality patterns." But creating a representative sample of biological systems seems infeasible. Rather than following this setup, we use a growing sample of biological systems and augment the pattern set with additional patterns if they are found. If the set of patterns remains stable throughout the analyses, we could claim that there is growing evidence that the pattern set is comprehensive.

The observations database includes 62 examples of problem–solution cases, biomimetic applications, and related biological systems. These examples were extracted from different biomimetic sources [5, 87, 129]. We chose examples of biomimetic innovations that their structure–function relations were studied and documented. The selected biological systems include most of the well-known reported biomimetic innovations in biomimetic databases and in the literature and in that manner, they fairly represent biological systems related to biomimetic innovations.

The observations database includes also 78 biological systems extracted from the same sources but without an application, and biological systems we encountered ourselves in our exercises. Though the biomimetic application does not exist, the challenge being solved by the biological system is clear. Altogether, our observations database includes 140 biological systems. It serves as the research database. If generic patterns do exist, we expect them to emerge from different data, no matter what data are observed. Therefore, if one would choose a different set of data, we expect him to recall the same patterns, at least part of them depending on the diversity of the data. Similarly, new patterns may emerge from different data in the future.

We analyzed biological systems derived from the observations database through a technical lens, using TRIZ tools and models, and searching for patterns. We identified three types of patterns; each type represents a different base of the structural biomimetic design method (Chaps. 7–9).

The analysis process itself is not only used as a research method but also integrated as part of the structural biomimetic design method, as presented in the method manual (Chap. 10). The manual corresponds with module 4 of our research model (Fig. 5.3), the new developed design method. The following chapters include the results:

Chapter 7: Base 1: Functional Patterns
Chapter 8: Base 2: Structure–Function Patterns
Chapter 9: Base 3: Sustainability patterns
Chapter 10: The structural biomimetic design method manual.

Chapter 7
Functional Patterns

7.1 The Analysis Process

We adapted the typical Su-Field analysis model to define the functions of the biological systems that appear in our observations database [172]. For each biological system we performed the Su-Field analysis, described in Table 7.1, by using the following steps:

- Identify system parts
- Identify system functions that can be explained as an interaction between two substances and a field.
- List functions that repeat in several groups of biological systems according to the patterns approach discussed in Sect. 6.2.

The analysis yielded biological function ontology, a minimal set of basic functions that characterize the biological domain and can be used in biomimetic design.

Remark Each system demonstrates various functions. This analysis may be done several times, each time observing a different function at the same system.

The analysis starts with scanning of possible system components by using the frame of system hierarchy presented in Sect. 6.1.1.2. As definitions of supersystem and subsystem depend on the context and functionality, we clarify that in our analysis, system boundaries are defined as the physical boundaries of the organism; the body casing. The system is the organism itself; subsystem parts are internal biotic organs or parts of the organism; supersystem parts are environmental elements outside the organism body casing which may be in the living surrounding of the organism. At this point, we didn't know which component will eventually be part of the Su-Field model. We then focused on one function and followed questions 3–5 in Table 7.1 to identify the target object, working unit and source of energy for their interaction. The working unit is in a physical contact with the target

Y.H. Cohen and Y. Reich, *Biomimetic Design Method for Innovation and Sustainability*, DOI 10.1007/978-3-319-33997-9_7

Table 7.1 The procedure stages for the Su-Field functional analysis

	Analysis procedure questions	Su-Field model components	Remarks
1	Identify system parts: system, supersystem, and subsystem	System parts	General mapping of system parts. Some of them will be identified as the Su-Field components
2	Identify the main function of the system	System observed function	
3	Which is the target object of this function?	Target object	
4	Which part of the system performs the function?	Working unit	The working unit is in a physical touch with the target object [155]
5	Which source of energy is used to perform the function?	Source of energy	

object according to the definition of a working unit [173]. Examples of this analysis for selected biological systems are presented in Table 7.2.

We repeated the Su-Field analysis procedure (Table 7.1) for each one of the 140 biological systems in our observations database and searched for functions that repeat in these analyses. We then classified functions that repeated in a large number of cases into high level classes and identified them as functional patterns or generic functions.

Table 7.2 Selected examples of Su-Field analysis of biological system

	Biological system function	Biological system name	Substance A (S_1)	Substance B (S_2)	Field (F)	Description of system interaction
1	Remove/add	Lotus leaf	Water droplet	Dirt particle	Intermolecular (adhesion)	Water droplets remove dirt particles from the leaf by adhesion forces as part of the self-cleaning mechanism of the lotus leaf
2	Contain	Bombardier beetle	Inner chamber	Noxious liquid	Mechanical (direct contact)	A small chamber less than one millimeter long within the beetle's body contains the noxious liquid as part of the defensive mechanism

(continued)

Table 7.2 (continued)

	Biological system function	Biological system name	Substance A (S_1)	Substance B (S_2)	Field (F)	Description of system interaction
3	Channel	Desert Rhubarb leaf	Ridged tunnels	Water droplets	Mechanical (gravitation)	The ridged tunnels of the desert rhubarb leaf channel the water droplets by gravitation forces
4	Regulate	Termite mound	Mound tunnels	Air particles	Thermal	Mound tunnels regulate air particles due to thermal ventilation
5	Push/stop	Sea snail shells	Potential predators	Shell layers	Mechanical (deformation)	The potential predators push the shell layers of the sea snail. The layers do not change permanently as they resist this mechanical pressure and prevent a deformation
6	Move/change	Samara Seeds	Wind (air particles)	Samara seed	Mechanical (wind)	Air particles (wind) move (rotate) the samara seed and change its location

7.2 Results

The functional patterns that repeat in our Su-Field analyses are named as the "Su-Field function ontology". They are presented in Table 7.3 aside other related functions and aside the basic interaction between the two substances that led to their definition.

Table 7.3 Su-Field function ontology

	Interaction between substance A and substance B	Su-Field function ontology	Related specified functions
1	A is added to B	Attach	Add, connect, attract, combine, increase
	A is removed from B	Detach	Remove, separate, subtract, cut, refract, decrease
2	A contains B	Contain	Hold, store, protect, defend, trap, grasp
3	A is channeled through B	Channel	Lead, stream, transmit, transfer, transport, guide, direct, flow
4	A is channeled through B on demand	Regulate	Control, modulate
5	A pushes B	Push	Push, press, pull, stress, crush
	B stops A	Stop	Push back, return, cease, secure, protect, isolate, insulate, stabilize, resist, smoothen, disperse, deflect
6	A changes the form of B	Change	Change form, transform, convert
	A changes the position/location of B		Turn, move up, move down, rotate, open, close, position

7.3 Discussion of Results, Explanations and Implications

Defining actions as a result of Su-Field analysis is usually expressed as a combination of one of the following verbs: increase, decrease, change and stabilize together with the name of the target object [136]. For example, the wind changes the position of the samara seed. We can see that these four basic verbs appear in Table 7.2: remove and add are equivalent to increase and decrease, stabilize appears as an elaboration of stop and change appears in the same wording. However, there are some other functions that do not appear in these usually expressed verbs such as contain, channel, push, stop and move that might characterize unique actions that are frequent in biological systems and not necessarily in technological systems.

The question why the results of the Su-Field function ontology are the functions described in Table 7.2 and not others may be explained by a morphological analysis of all the possible interactions between two substances in a given space. This analysis corresponds to Zwicky's morphological approach [174] to analyze logically what could be all the options that span the space where our phenomena lies.

$$\mathrm{A+B} \quad \text{(A is attached to B)}$$

$$\mathrm{A-B} \quad \text{(B is detached from A)}$$

$$\mathrm{A \subset B} \quad \text{(A is entered into B)} \rightarrow \text{entered permanently (contained)}$$
$$\rightarrow \text{entered temporarily (channeled, regulated)}$$

$$\mathrm{A \rightarrow B}\text{(A pushes B, B pushes back A) with no permanent change in B}$$

$$\text{A changes B} \rightarrow \text{A changes the form, position or place of B}$$

For example, suppose we pour water (substance A) on a non-permeable surface (substance B). Under a field of adhesion forces, the water may be attached to the surface creating a new distinguished unit of water plus surface. Under a field of gravitation, if the surface is positioned in a slope, the water may be detached from the surface, and the result is a separation of these two substances. Now, let's assume the surface is curved and has a tube shape. In this case, the water will channel through the surface under a certain field of gravitation or capillarity forces. But, if the curved surface has a glass shape, if we pour the water into the surface, the surface will contain them as the surface is not permeable. The water pushes the glass and the glass pushes back the water. As a result, the surface does not change permanently. But if the surface is made of a very thin and weak layer of material, the water may push the surface and move it to a new location or cause it to crack or fracture. If we substitute the surface material with a permeable material like a surface of sand, than the result is a new function, mix. As we analyzed biological organisms on the system levels we didn't find this function, which is more frequent in chemical processes. This morphological analysis demonstrates the space of possible interactions between two substances and one field. As demonstrated, if we change the field or some of the substance characteristics such as shape, position, permeability or thickness, we get different interactions or different functions.

This morphological analysis introduces a limited list of possible interactions (actions) between two substances in a given space. The definition of functions according to this morphological analysis is similar to the functions derived by the Su-Field analysis and represents possible interactions between two substances in a given space.

7.3.1 Comparing Su-Field Ontology to Selected References

Several functional ontologies were developed as a result of the need to support functional modeling processes. In this section, we compare our Su-field function ontology, derived from biological systems, to selected references of previous engineering functional ontologies. The results of this comparison appear in Table 7.4.

Pahl and Beitz [94] used the approach of associating functions to material, energy and signal flows and identified five general functions: Channel, Connect, Vary, Change and Store. Hundal [175] listed six function classes: Branch, Channel, Connect, Change Magnitude, Convert and Store/Supply. For each class he related more specific functions such as valve, stop and crush. Altshuller [132] offered a set of 30 functional descriptions which may be distilled into less functions according to Pahl and Beitz taxonomy: Separate, Transfer, Change, Accumulate, Control and Stabilize. Hirts et al. [138] aimed to generate an atomic taxonomy that is generic enough to be used for modeling a broad variety of engineering products. They developed a consistent functional vocabulary by integrating previous research efforts and came up with the functional basis of the following functions: Branch,

Table 7.4 Comparison between Su-Field function ontology to previous engineering functional ontologies (adapted from Hirtz et al. [138, p. 69])

	Su-Field function ontology	Pahl and Beitz	Hundel	Altshuller	Hirtz
1	Attach	Connect	Connect	–	Connect
	Detach	–	Branch (separate, cut)	Separate	Branch (separate, remove)
2	Contain	Store	Store/supply	Accumulate	Provision (store, supply)
3	Channel	Channel	Channel	Transfer	Channel
4	Regulate	Vary	Valve	Control	Control magnitude, regulate
5	Push	–	Crush	–	–
	Stop	–	Stop	Stabilize	Stop
6	Change	Change	Convert, switch, mix	Change	Convert

Channel, Connect, Control Magnitude, Convert, Provision, Signal and Support (stop).

Su-Field analysis of biological systems (Tables 7.3) yielded a similar ontology to previous engineering functions ontologies, and therefore it may be useful to mechanical engineers who are familiar with other functional bases. It contains the core functions that appear in all compared ontologies: Attach/Detach (Connect/Separate), Contain (Store), Channel, Regulate and Change. It also contains other functions like Push and Stop that appear in some of the compared ontologies. According to this comparison we may suggest that Su-Field function ontology is a generalization of system function ontologies, as it provides a minimal set of functions that encompass the previous function ontologies. However, functions related to processes such as Signal, were not identified by the Su-Field analysis that focused on analyzing structured based phenomena.

7.3.2 Summary

An analysis of dozens of biological systems by Su-Field model led to a definition of biological function ontology (Table 7.3). This ontology encompasses functions that repeat in large numbers of cases in our observations database. All the functional mechanisms may be explained by an interaction of two substances and one field of energy, though the substances and the field changes between the cases. A comparison of the Su-Field ontology to previous engineered ontologies (Table 7.4) showed that Su-Field ontology is an efficient generalization of system function ontologies.

The results have implications for the field of functional modeling in general, and especially for the field of interdisciplinary design, such as biomimetic design.

Su-Field function ontology may be a top-level function ontology that supports functional modeling processes as it is clear, simple and generalize other function ontologies. Regarding biomimetic design processes, Su-Field function ontology may be the bridge between biology and technology as function is considered to be the linking concept between these two domains, in a process that is known as the "function-bridge" [5]. Defining the design challenge in terms of the required function enables to locate a suitable model for inspiration or imitation in nature that achieves the same function. Therefore, providing a minimal comprehensive set of biological functions that repeat in biological systems and have similar meaning and wording in technological systems, may assist designers to bridge the gap between biology and technology.

Chapter 8
Structure-Function Patterns

8.1 The Analysis Process

We adapted the law of system completeness to analyse the structure of the biological systems that appear in our observations database. Following the Su-Field analysis that already identified the working unit of the law of system completeness, we elaborated the analysis to include the other parts of the model according to the stages described in Table 8.1. Thus we identified in addition the engine, the transmission unit and the control unit of the observed biological system [176]. Altogether, the analysis of each biological system included both parts of Su-Field and law of system completeness analyses, thus providing a full complete viable system model, as presented in Table 8.1. We then classified structures that repeated in a large number of cases into high level classes. We revealed that these structures correlate with the same generic functions presented before as the Su-Field function ontology (Table 7.3), hence, identified them as structure-function patterns. Through this approach we analysed 140 biological systems, and identified the existence of a set of structure-function patterns. The logic for searching for these groups of repeated structures/structure-function relations is the patterns approach discussed in Sect. 6.2.

8.1.1 Analysis Examples

8.1.1.1 The Lotus Leaf Cleaning System

Description of the mechanism: Dirt particle removal by a water droplet is a stepwise process. There is a need to attach the dirt particle to the water droplet and then move the water droplet accompanied with the dirt particle away from the leaf. In this example we focus on the first stage of detaching the dirt particle from the leaf.

Y.H. Cohen and Y. Reich, *Biomimetic Design Method for Innovation and Sustainability*, DOI 10.1007/978-3-319-33997-9_8

Table 8.1 The procedure stages for the law of systems completeness analysis

	Analysis procedure questions	The law of system completeness parts
1	Which part of the system converts the source of energy (field) to work?	Engine
2	Which part of the system transfers this energy to the working unit?	Transmission unit
3	Which part of the system initiate or stop the action?	Control unit

The Lotus leaf (Nelumno nucifera) structure is full of small epidermal protrusions covered with wax at the nanometer range (see Fig. 8.1). These protrusions create high contact angle between the surface and the droplet resulting in a reduced contact area and a reduced adhesion force between the surface and the droplet. The droplet gets a spherical shape and rolls down the slope of the leaf under the influence of gravity. If it rolls across a dirt particle, the droplet collects it, as the adhesion between the dirt particle and the droplet is higher than the adhesion between the dirt particle and the leaf surface [53]. This self-cleaning mechanism is considered as the lotus effect.

The analysis stages for this example are described in Table 8.2 (under engine model). The resulting system diagram is shown in Fig. 8.2. The bold lines are components of the Su-Field analysis. The dashed lines are components of the law of system completeness. Together the diagram is adjusted to the complete viable system model described in Sect. 6.1.3.3. Each box is denoted with its role in the system. This template is used in subsequent examples.

As shown in Fig. 8.2, the water droplets, environmental elements, perform the work of moving the particles. Control is done by environmental conditions as the work is based on the presence of water and particles in the environment. It is clear that both the working unit and the control are done by environmental components that are external to the leaf system boundaries. It is also clear that the system engine, identified as the epidermal protrusions, is exploiting available energy sources as adhesion and gravitation fields.

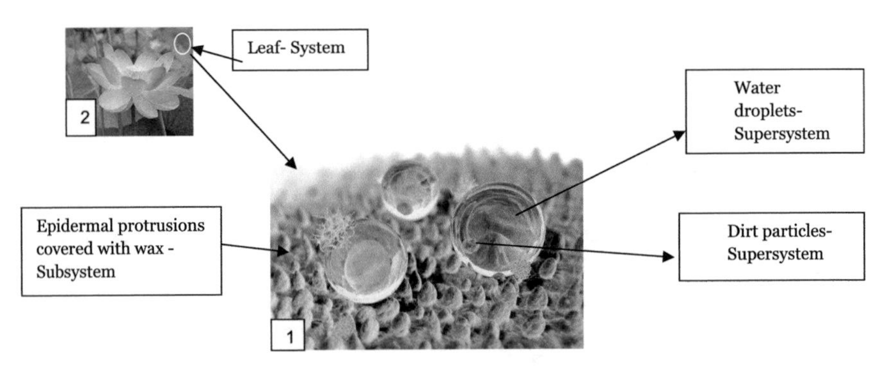

Fig. 8.1 The lotus leaf cleaning mechanism. *Photo 1* by William Thielicke from Wikimedia under GNU Free Documentation License, Version 1.2. *Photo 2* by Peripitus adapted from Wikimedia under GNU Free Documentation License, Version 1.2

Table 8.2 The procedure stages for the complete viable system analysis

		Analysis procedure questions—Engine model	Analysis procedure questions—Brake model	Complete viable system component	Remarks
1	System parts hierarchy	Identify system parts: System, Supersystem, Subsystem	Identify system parts: System, Supersystem, Subsystem	System parts	–
2	Su-Field analysis	Identify the main function (action) of the system	Identify the main harmful function (action)	System observed function	–
3		What is the target object of this function?	What is the target object of this harmful function?	Target object	–
4		Which part of the system performs the function?	Which part of the system performs the harmful action?	Working unit	The working unit is in physical touch with the target object
5		Which source of energy is used to perform the function?	Which source of energy is blocked to prevent the harmful action?	Energy source	–
6	The law of system completeness model	Which part of the system converts the source of energy (field) to work?	Which part of the system stops this source of energy?	Engine/Brake	–
7		Which part of the system transfers this energy to the working unit?		Transmission unit	Taken out from model
8		Which part of the system initiate or stop the action?	Which part of the system initiate or stop the harmful action?	Control unit	–

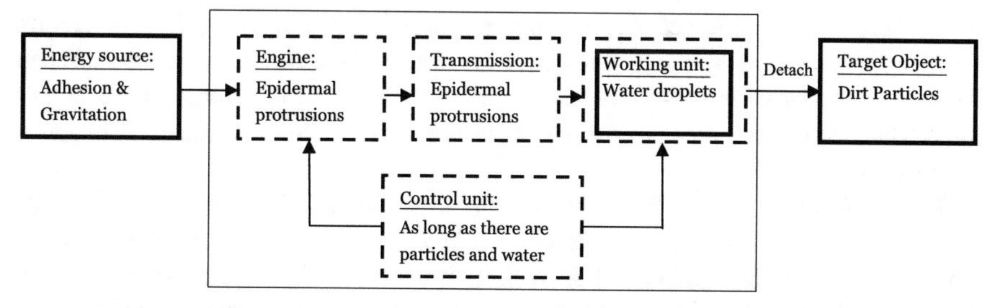

Fig. 8.2 Lotus leaf cleaning mechanism—analysis by the complete viable system model

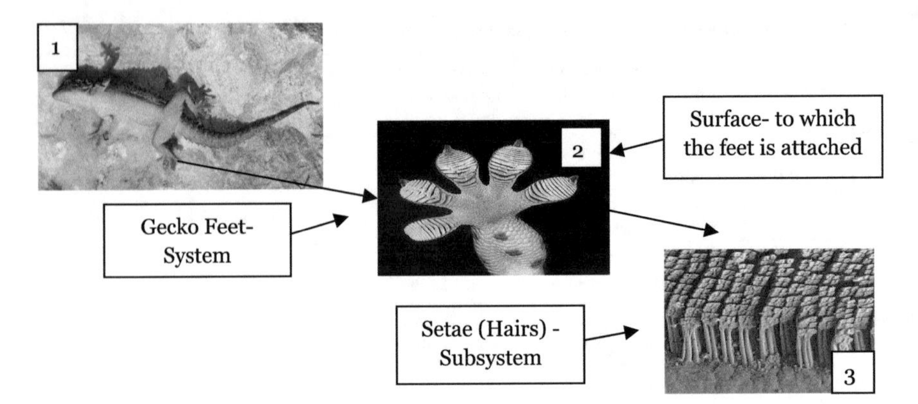

Fig. 8.3 The gecko's feet adhesion mechanism. *Photo 1* by Yanpetro adapted from wikimedia under GNU Free Documentation License, Version 1.2. *Photo 2–3* by Kellar Autumn [177], reproduced with permission of the copyright owner, Kellar Autumn

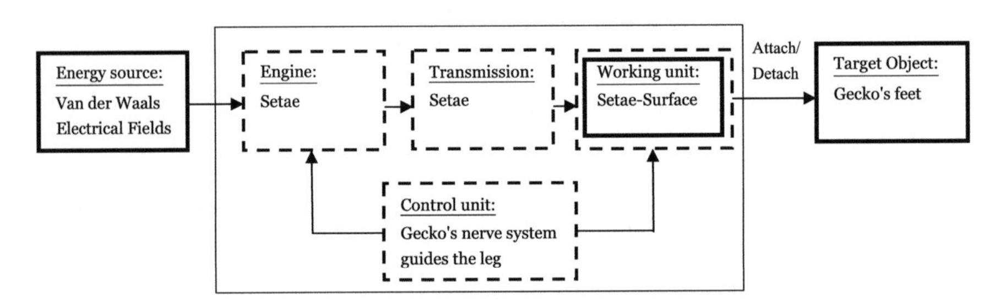

Fig. 8.4 Gecko's foot adhesion mechanism—analysis by the complete viable system model

8.1.1.2 The Gecko Feet Attachment/Detachment System

Description of mechanism: Gecko's foot has nearly five hundred thousand hairs or setae (see Fig. 8.3). Each individual setae is connected to the surface by van der Waals forces as discovered by Autumn et al. [177]. Multiplying the connection force of each setae by the total numbers of setae creates a strong connection. When the gecko changes the foot's contact angle with the surface, these van der Waals forces disappear. The foot is switched between two situations: Attached/Detached to the surface.

The analysis stages for this example are described in Table 8.2 (under engine model). The resulting system diagram is shown in Fig. 8.4. It is clear that the working unit includes the interaction with the surface that is a supersystem element, external to the gecko's feet boundaries. In this case, control is done by the gecko's nerves system that guides the leg against and from the surface. It is also clear that the system engine, identified as the setae, is exploiting available energy sources, identified as the electrical field formed by the van der Waals forces.

8.1.1.3 The Click Beetle Jumping System

Description of Mechanism: The click beetle has large longitudinal muscles in its body (see Fig. 8.5). These muscles are divided into two subunits by a hinge made of cuticular peg. The muscles contract to store elastic energy. When the peg slides and frees the hinge, the stored energy is abruptly released; then, the body is flexed and lifted from the ground within less than 1 ms [178].

The analysis stages for this example are described in Table 8.2 (under engine model). The resulting system diagram is shown in Fig. 8.6. As shown in Fig. 8.6, the contracted muscles serve as the engine, transmission, and the working unit. The elastic energy is stored in the muscles and is used to lift the beetle body when the hinge is released. The control unit is the beetle's nerve system.

8.1.2 *The Analysis Stages*

See Table 8.2.

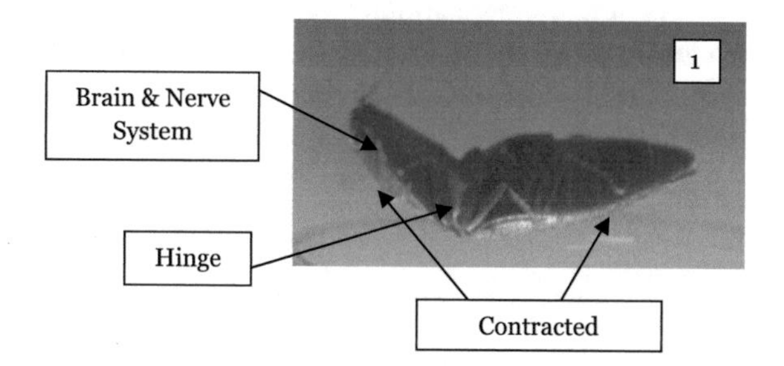

Fig. 8.5 The click beetle jumping mechanism. *Photo 1* by Gal Ribak reproduced with permission of the copyright owner

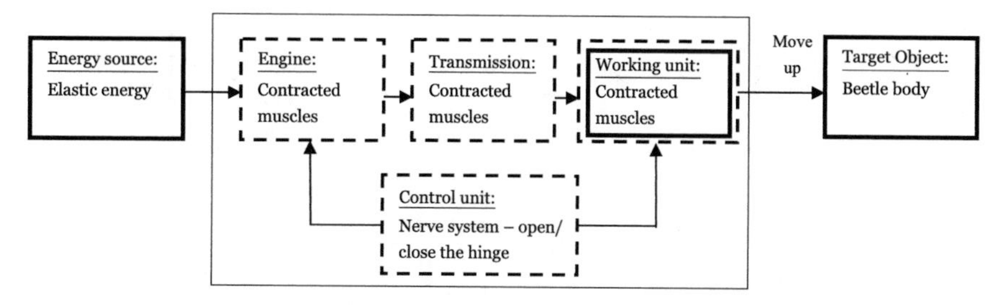

Fig. 8.6 The click beetle jumping mechanism—analysis by the complete viable system model

8.2 Results—The Complete Viable Model

8.2.1 Transmission Unit

We did not identify the transmission unit at the level of our analysis. It might exist at different levels such as the cell level. As we intend to use this complete viable system model at the organism or organ level, we suggest taking out the transmission unit and simplify the model as presented in Fig. 8.7. The bold solid lines show the system parts that belong to the Su-Field model while the dashed lines depict the system parts that belong to the law of system completeness model.

8.2.2 Engines and Brakes

Some of analyzed systems did not match the classical complete viable system model. In these cases, the fields of energy where not exploited by the engines but rejected from the system. As a result, we defined a new component, a brake, replacing the engine component in these cases. The brake blocks energy field and prevents the formation of undesired harmful function. Accordingly, another version of the complete viable system model is suggested and presented in Fig. 8.8. The bold solid lines show the system parts that belong to the Su-Field model while the dashed lines depict the system parts that belong to the law of system completeness model. Further elaboration on 'engine' and 'brake' definitions is presented in Sect. 8.4.

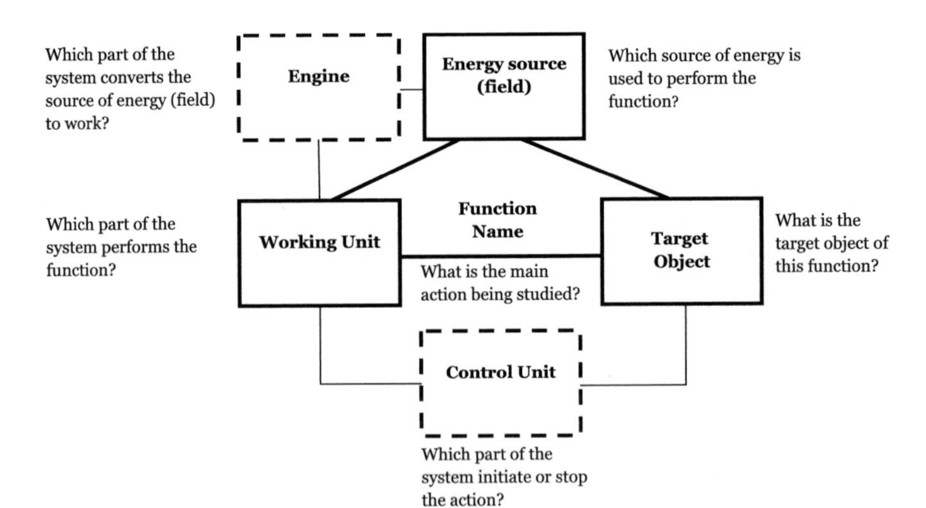

Fig. 8.7 The complete viable system simplified engine model

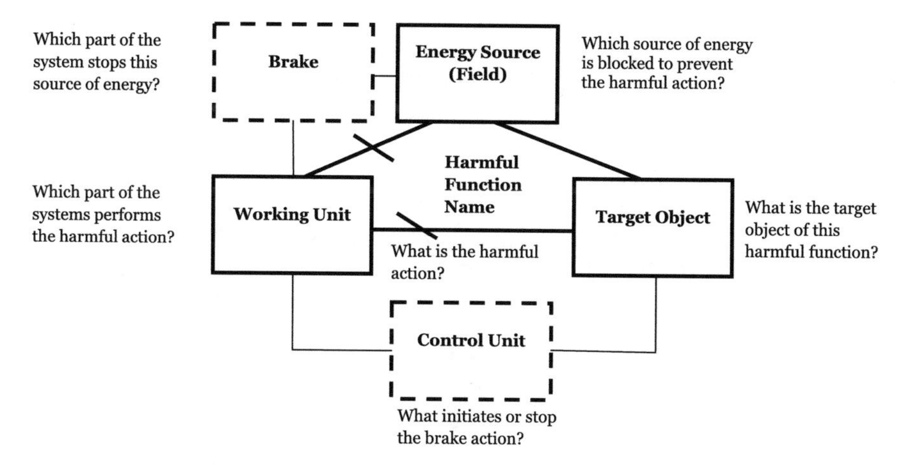

Fig. 8.8 The complete viable system simplified brake model

8.2.2.1 An Example of a System Analysis with a Brake

Description of mechanism: The shell of the abalone is combined of alternating hard layers made of microscopic calcium carbonate and soft layers made of protein substance (see Fig. 8.9). It is a perfect example of a composite material. When the abalone shell is exposed to external pressure, the hard layers slide instead of breaking and the protein stretches to absorb the energy of the pressure. The protein acts like "rubber" and has enormous capacity to absorb shock without breaking [179].

The analysis stages for this example are described in Table 8.2 (under brake model). The resulting system diagram is shown in Fig. 8.10. The soft layers made of protein act as the brake of the system and block the mechanical energy of external pressure, such as potential predator. The brake defends the abalone shell

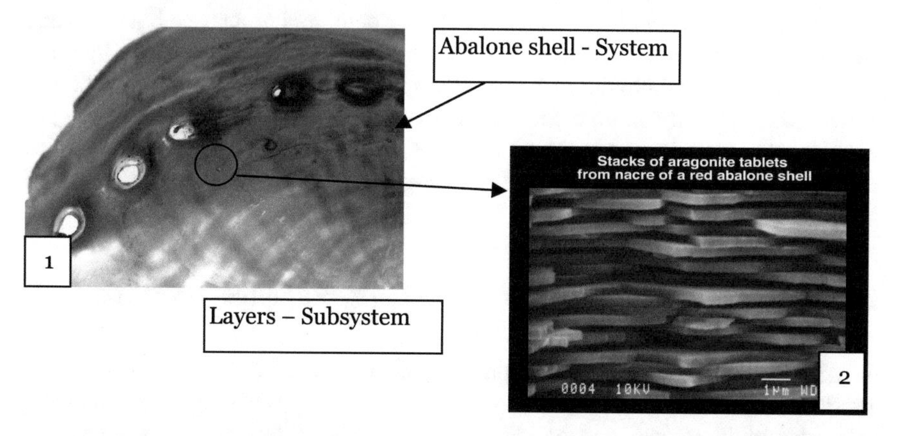

Fig. 8.9 The abalone shell pressure absorption mechanism. *Photo 1* by N Yotaro adapted from Wikimedia under GNU Free Documentation License, Version 1.2. *Photo 2* by Paul Hansma [179], reproduced with permission of the copyright owner

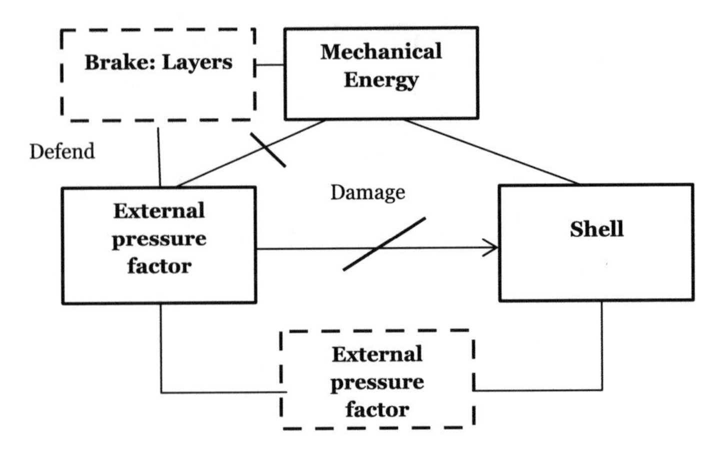

Fig. 8.10 The abalone shell pressure absorption mechanism—analysis by the by the complete viable system model with a brake

and prevents a potential damage to the shell. The braking process happens with the presence of external pressure.

8.2.3 The Complete Viable System Model as an Abstraction Tool

We demonstrated the applicability of the complete viable system model in the domain of biological systems, a new and uninvestigated domain in relation to this model as far as we know. A designer may follow Table 8.2 (and skip stage 7—the transmission unit taken out of the model). The complete viable system model accompanied with the analysis process (Table 8.2) is offered as a modeling tool to analyse both the functional and the structural aspects of a biological system. This tool can be further used for analysing other biological systems through a technical lens for biomimetic purposes. Analogy to the biomimetic application target can be easily abstracted as presented in the manual section. The advantage of this tool is that it is based on a technical view of a biological system, a view that an engineer may feel more comfortable with.

8.3 Results—Sustainability Aspects of Biological Systems

In addition to the structure-function patterns we revealed by this analysis and present in Sect. 8.4, the analysis also enriched our understanding of sustainability strategies of biological systems. Specifically, we identified several sustainability design strategies in nature including:

- Utilizing environmental energy sources—Biological systems use external energy sources almost without any extra cost by adopting readily available environmental resources. The engine components of the biological system are adjusted to exploit available energy sources.
- Utilizing environmental elements as essential systems parts—In biological systems, essential elements, such as the working unit or the control unit, may be provided by elements and components from their surrounding environment, meaning they are external to the biological physical boundaries.
- Unification of system parts—Multifunctional design is one of the prominent design principles in nature, providing several functions by a single component. We demonstrated a tendency in some biological systems to combine several essential parts of the systems to the same component, mainly the engine and the transmission unit, but sometimes the working unit as well.

Sustainability design principles are further discussed in Chap. 9.

8.4 Results—Structure-Function Patterns

According to the previously presented analysis of the biological systems by the complete viable system model (Figs. 8.2, 8.4, 8.6 and 8.10), we repeated the analysis for each biological system in our observations database. We identified which structures are the model components (engines, brakes, working unit, control unit and target object) in relation to an observed function and what is the source of energy for that action. We then classified structures that repeated in a large number of cases into high level classes, revealed that they correlate to the same functions derived from the Su-Field function ontology (Table 7.3), and identified them as structure-function patterns.

The analysis identified several structure-function patterns at the meta level that mainly repeat in existing biomimetic applications [180]. Some of the structures were identified as the engines of the system according to the law of system completeness. In these cases, fields of energy are used by the engine structures to perform the required function. In other cases, the structures block fields of energy, when these are harmful to the system. We identified them as brakes, though a brake is not part of the original law of system completeness model.

Man-made engines are machines designed to convert energy into motion. Similarly, engines in nature are structures that exploit the potential propulsion properties of forces in space as clean and renewable energy sources for the purpose of performing a motion. Their correlated main functions are dynamic and are related to motion: to channel, to change, to attach, and to detach. Generally, engines are related to the function 'Move'.

A brake is a device that inhibits motion. Similarly, brakes in nature are structures that inhibit undesired motion resulting from forces in space. They sustain different loads by absorbing the loads, pushing them back or changing their direction. Nature structures that inhibit forces from causing an undesired motion may be identified as brakes. Their correlated main functions are static and are related to state of no motion: to protect, to defend, to stabilize. Generally, brakes are related to the function 'Stop'.

A list of the repeated structure-function patterns and their classifications as engines or brakes is summarized in Table 10.4. The following sections provide examples for each structure-function pattern including its role as an engine that uses external fields of energy or a brake that blocks them.

8.4.1 Engines

8.4.1.1 Repeated Protrusions

This structure is identified with repeated protrusions of different materials and shapes, at the nanoscale or at the macroscale. Protrusions may be hairs, denticles, bristles, epidermal protrusions, scales and more. Examples and images are provided in Table 8.3. These protrusions are correlated to generic functions of 'Attach' or 'Detach' to different target objects such as dirt particles, bacteria and water droplets. They are also associated with attachment to different sorts of surfaces like walls, sand, soil, ice, fur and more. The repeated protrusion structure exploits space forces or gradients in order to perform the attachment or detachment function. The repetition of protrusions enlarges the surface area of the structure and intensifies the interaction of the structure with the forces.

8.4.1.2 Repeated Tubes/Channels/Tunnels

This structure is identified with repeated tubes or channels with or without valves. It appears at different scales. Examples and images are provided in Table 8.4. The structure is correlated to generic functions of 'Lead' or 'Channel' if it has no valves or to the function of 'Regulate' if it has valves. The repeated tube structure exploits force gradients to channel target objects or to regulate their flow. The repetition of tubes increases the surface area that is being interacted within the environment and thus intensifies the performed function.

8.4.1.3 Asymmetry

Although symmetric structures are common in nature, asymmetric structures also prevail. **Geometric asymmetry** is defined when the structure has at least one dimension without symmetry. **Time asymmetry** is defined when the structure demonstrates different appearances at different times. **Material asymmetry** is defined when the material properties of a structure are distributed asymmetrically in

Table 8.3 Repeated protrusions structures: Examples and images

	Organism	Repeated protrusions	Function	Image
1	Gecko's foot	Repeated hairs—Gecko's foot has nearly five hundred thousand hairs (setae). Each individual setae is attached to and detached from the surface by van der Waals forces [177] due to electrical gradient. As a result, the gecko's foot is attached and detached from the surface	Attach/Detach foot	
2	Lotus leaf	Repeated Epidermal protrusions—the Lotus leaf (Nelumno nucifera) is full of small epidermal protrusions covered with wax at the nanometer range. These protrusions create high contact angles between the surface and the droplets resulting in an adhesion gradient that together with the gravitation gradient detach dirt particles from the leaf and roll them away [53]	Remove Dirt Particles	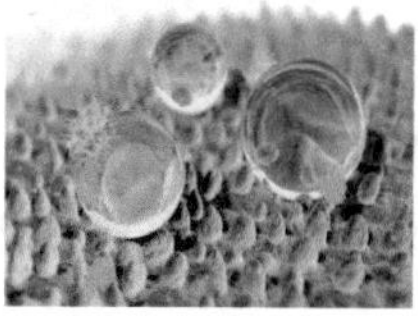
3	Shark skin	Repeated Denticles—The shark's skin has dermal denticles (little skin teeth) which are ribbed with longitudinal grooves. This structure is related to removal of fouling and bacteria. The denticle's surface tension characteristics are used to remove the fouling [187]	Remove Bacteria	
4	Burdock plant	Repeated hooks—The burrs of the burdock plant are covered with tiny hooks which are adjusted to be connected to the animal's fur and may exploit their movement ability to spread away	Connect to animal fur	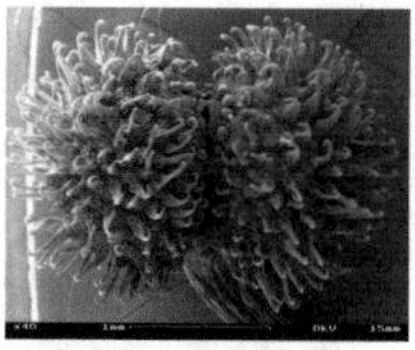

Photo 1 by Kellar Autumn [177], reproduced with permission of the copyright owner
Photo 2 by William Thielicke from Wikimedia under GNU Free Documentation License, Version 1.2
Photo 3 by Sharklet Technologies, reproduced with permission of the copyright owner
Photo 4 by Secret Disc from Wikimedia under GNU Free Documentation License, Version 3.0

Table 8.4 Repeated tubes/channels/tunnels structures: Examples and images

	Organism	Repeated tubes/Channels/Tunnels	Function	Image
1	Termite Mound	Termite mounds are constructed of an extensive network of tunnels. The tunnels structure is considered to be an analogue of a lung, responsible for the global function of colony gas exchange. The regulation of gas is a complicated process that is related partially to ventilation inside the mound tunnels. In the egress tunnels air movement are driven by wind [22]	Regulates the gas exchange	
2	Tree	At the macroscopic level, wood is composed of parallel hollow tubes formed by the wood cells shape. These tubes are responsible for leading water and nutrients from the ground towards the tree's organs due to capillary forces opposing gravity [184]	Lead water and nutrients	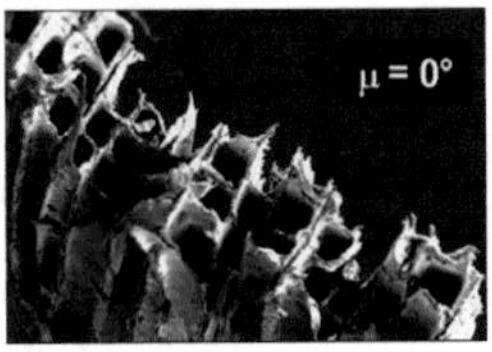
3	Cell membrane	The membrane of a cell is constructed of two layers of lipid cells. Between these two layers there are channels of proteins which allow a passive movement (diffusion) of ions (ion channels), water (aquaporins) or other solutes through the membrane down their electrochemical gradient	Regulate ions and water levels	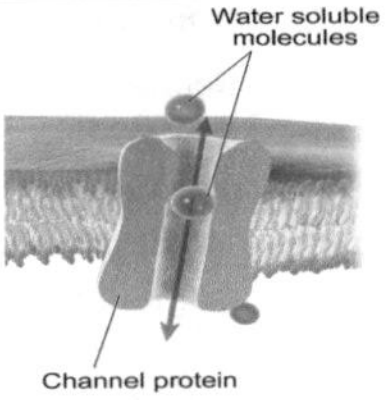
4	Desert Rhubarb	The desert rhubarb has large leaves with a ridged structure that creates a surface of channels. These channels lead the rain towards the ground near the plant's root by the gravitation gradient [188]	Channel water	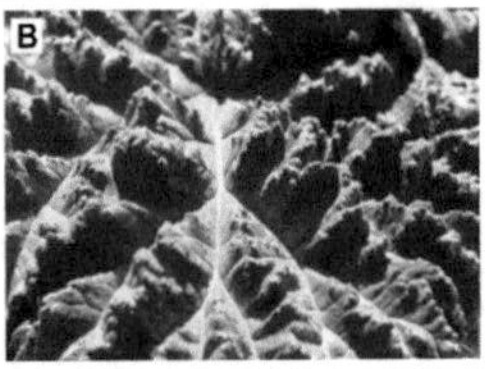

Photo 1 by Rupert Soar, reproduced with permission of the copyright owner
Photo 2 by Peter Fratzl [184], reproduced with permission of the copyright owner
Photo 3 by Blausen.com staff. "Blausen gallery 2014". Wikiversity Journal of Medicine
Photo 4 by Simcha Lev-Yadun [188], reproduced with permission of the copyright owner

time or space. This asymmetry creates variant reactions to external factors. In geometric asymmetry, different shapes react differently to external forces. In material asymmetry, different materials react differently to external factors like levels of humidity, light, temperature and mechanical pressures. Asymmetric structures are associated with generic functions of 'Change'. Variant reactions to external factors create gradients that are exploited to change the position or form of the structure. Examples are provided in Table 8.5.

Table 8.5 Asymmetric structures: Examples and images

		An Example	Function	Image
1	Geometric Asymmetry	The basking shark swims with open mouth in order to catch its food. Its stretched jaw demonstrates an asymmetric structure as the inner part of the jaw is straight and the outer part of the jaw is curved. This asymmetric structure creates a passive flow of water through the shark's gills due to the pressure gradient between the bottom and the upper parts of the jaw. Thus, excess water are moved without investing precious energy or structural complexity compared to a solution with a designated pumping organ [189]	Move up water	
3	Material Asymmetry	Pine scales are combined of two tissues. The inner tissue has a significantly lower coefficient of hygroscopic expansion of fibers than the outer tissue. As a result the scales are closed when it is humid and opened when it is dry [190]	open and close scales	
2	Time Asymmetry	Puffer fish (Tetraodontidae family) is able to inflate its body by swallowing water. The increased size of the inflated body makes it more difficult to be swallowed by predators. Thus the body volume is changed on time [191]	Change volume of fish body	

Photo 1 Photo from public domain
Photo 2 by Fir0002 from Wikimedia under GNU Free Documentation License, Version 1.2
Photo 3 by Brocken InaGlory from Wikimedia under GNU Free Documentation License, Version 1.2

8.4.2 Brakes

8.4.2.1 Mechanical Structures

The following structures are associated with the generic function of 'Protect' or 'Defend' from mechanical loads. Some of these structures may be effective also in exposure to thermal or chemical loads. Each one of them is adjusted to different types and directions of loads. Examples are provided in Table 8.6.

Table 8.6 Mechanical structures: Examples and images

		An Example	Function	Image
1	Repeated Layers	The shell of the abalone combines alternating hard layers made of brick-like units of calcium carbonate and soft layers made of protein substance. When the abalone shell is exposed to external pressure, the hard layers slide instead of breaking and the protein stretches to absorb the energy of the pressure. The protein acts like "rubber" and has enormous capacity to absorb shock without breaking [179]	Protect the abalone (Absorb pressure)	Stacks of aragonite tablets from nacre of a red abalone shell
2	Intersected layers	The toucan beak is constructed of an external solid keratin layer and internal fibrous network of a closed cell foam-like structure made of struts, which together with protein membranes, enclose shaped air spaces. The foam stabilizes the deformation of the beak by providing an elastic foundation which increases its buckling load under flexure loading [183]	Protect the beak	
3	Hollow Cylinder/tube	A human bone is a hollow tube which resists the tendency to break [160]. As the resistance to a load is proportional to the thickness of the structure in the direction of the load, the material is added when required (at the joints) and removed when it is not required (at the midpoint) [160]	Protect the bone	
4	Helical structure	The intervertebral disc is composed of helically collagen fibers that work like shock absorber and protect the disc from torsion, flexure and extension forces [192]	Protect the disc	

Photo 1 by Paul Hansma, reproduced with permission of the copyright owner
Photo 2 by Seki. reproduced with permission of the Meyer's group [183]
Photo 3 by Peter Fratzl, reproduced with permission of the copyright owner [184]
Photo 4 by Deborah Spurlock, Indiana University Southeast, reproduced with permission of the copyright owner

8.4.2.2 Repeated Layers—Sandwich Structure

A Sandwich structure is a composite combined of several parallel alternating layers of different materials with different properties. This structure resists bending in one particular direction and is strong, stable, tensile, durable and temperature resistant [181]. This structure sustains the mechanical load by absorbing it in the loose or soft layers and distributes it over large areas of the structure surface [182]. In case of thermal pressures, this structure provides a protective cover and insulation that minimizes heat transfer. In this case, this structure consists of layers of hairs or feathers, fat and air, which are poor heat conductors [182].

8.4.2.3 Intersected Layers—Crisscrossed Structure

This structure consists of a cross linking of fibers that create a crisscrossed structure. In two dimensions it looks like a network, in three dimensions it may look like foam, a cellular structure or honeycomb structure. This structure is common in beaks of birds, bones, shells of lizards and insect wings. This structure provides an elastic foundation with high resistance to flexure loading [183]. When fixed specified loading direction exist, like gravity, this structure exhibits more efficient use of its material [184]. In case of mechanical pressures, this structure absorbs the mechanical loads within the spaces between the fibers.

8.4.2.4 Hollow Cylinders (Tubular) Structure

The structure is characterized with a hollow cylinder, rod or tube. The hollow cylinder appears in the quill of the bird's feather, stems of flowers, bamboos and reeds, stalks of grain, limbs of insects, tree cells and long human bones, like the femur. Most houses of ground dwellers have a tubular design. The hollow cylinder provides stability against bending and buckling [184] and is adjusted to resist bending in all directions [185]. We note that in the case of repeated tubes structure, each tube in the arrangement functions as a single tube.

8.4.2.5 Helical Structure

The structure is characterized by a helical form in three dimensions. Helical forms appear in crystalline materials, seed pods and plant tendrils. They are the base structure of the DNA and appear frequently in the human body [185]. The helical structure is associated with mechanical efficiency. It stores and absorbs energy, provides flexibility and stability and prevents wrinkling [186]. It resists circumferential and longitudinal stresses while an optimum fiber angle of 55 (axial) balances both these stresses [185]. An array of helical fibers under pressure resists torsion deformation and bends smoothly without kinking. Compared to a crisscrossing pattern, which is not effective in withstanding inner pressures, the helix structure is effective in this case [160].

8.4.2.6 Streamlined Structures/Shapes

Streamlined shapes are contours designed to minimize resistance to motion through a fluid (such as air). Streamlined geometries are associated with efficient flow. They minimize kinetic energy loss to heat through friction as they change the direction of the flow and create laminar instead of turbulent flow. They are associated with the generic function of 'Stabilize' the target object in the presence of fluids flows.

One common streamlined structure in nature is the spiral (Fig. 8.11a). The efficient flight and swimming patterns in nature are based on variations of spiral flows. Other examples are the penguins body contour that is related to their fast and efficient swim (Fig. 8.11b), the boxfish streamlined body that excels in stable, smooth and energy-efficient swimming (Fig. 8.11c) and the beak of the kingfisher that enables the kingfisher to travel smoothly between the medium of air and water, without losing energy to noise (Fig. 8.11d). Streamlined shapes appear also in non-animate systems like river basins [8].

8.4.2.7 Container Structure

Container structure is an enclosed cavity encompassed by an external cover. The cavity may be full of gases, liquids or solids. Its generic function is to 'Contain'. It is adjusted to sustain the load of gravity by exerting resisting load thus inhibiting the motion of the contained object as a result of gravitation. In some cases it also sustains external mechanical loads.

One example of a container structure is the sphere which is a basic form in nature that has the ability to absorb and dissipate loads due to its curvilinear form [162]. Other examples are bird nests that hold and protect the eggs (Fig. 8.12a), burrows of animals such as the Jaw fish that contain and protect the animals against predators (Fig. 8.12b), geophytes that serve as underground storage organs (Fig. 8.12c) and carnivorous plant cups that contain digestive enzymes (Fig. 8.12d).

Fig. 8.11 Streamlined shapes. **a** Spiral shell. **b** Penguins body contour. **c** Boxfish. **d** Kingfisher's beak. *Photo* **a** by Andrew Butko from Wikimedia under GNU Free Documentation License, Version 1.3. *Photo* **b** and **c** from public domain. *Photo* **d** from Biomimicry IL, with permission

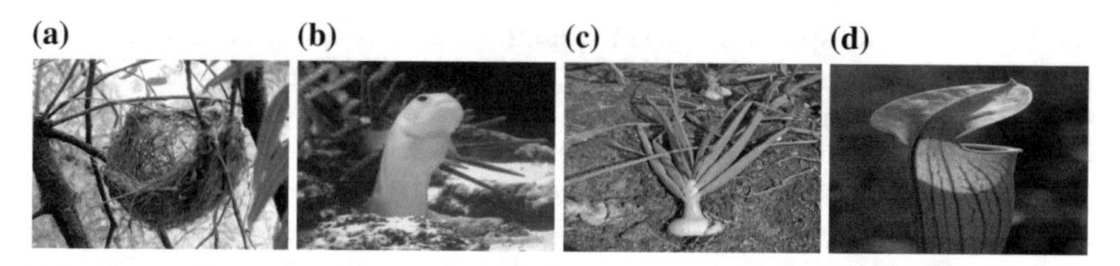

Fig. 8.12 Container structures. **a** Bird nests. **b** Jaw fish-burrow. **c** Geophytes. **d** Carnivorous plants cups. *Photo* **a** by Fir0002 from Wikimedia under GNU Free Documentation License, Version 1.2. *Photo* **b** from Wikimedia under GNU Free Documentation License, Version 1.2. *Photo* **c** by H. Zell from Wikimedia under GNU Free Documentation License, Version 1.2. *Photo* **d** from public domain

8.4.3 Statistics and Frequency of Patterns Occurrence

We analyzed 140 biological systems that altogether yielded 193 patterns with the breakdown shown in Table 8.7. These statistics reflect also the frequency of functions, as each structural pattern is related to one generic function. In most cases, the relation of structure to function in our database is one-to-one with exception of the tube and muscles that are dual structures serving both as engine and brake. Altogether, these do not change the statistics much.

As most patterns are derived from biomimetic sources, their frequency in our database does not reflect their general frequency in nature. It is interesting to realize that repeated protrusions structure is quite frequent in biomimetic applications. The reason for that may be its visual prominence that might capture the biomimetic designer attention. It is related to generic function of 'Attach' or 'Detach' that are the core of many biomimetic applications. Among the biological systems that demonstrate more than one pattern we can identify some frequent combinations of patterns that go together. Repeated protrusions tend to appear with repeated

Table 8.7 Statistics of patterns appearance in our database

		Structural pattern	Frequency of occurrence	Summary frequency of occurrence
Engines	1	Repeated protrusions	45	109
	2	Repeated tubes/Channels/Tunnels	29	
	3	Asymmetry	35	
Brakes	4	Repeated layers—Sandwich structure	26	84
	5	Intersected layers—Crisscrossed structure	10	
	6	Hollow cylinders (tubular) structure	10	
	7	Helical structure	5	
	8	Streamlined structures/Shapes	12	
	9	Container structure	21	
Total				193

channels, when the spaces between the protrusions have a channel shape and are used to channel water. Container structures tend to appear with repeated or intersected layers when there is a request to provide protection for the stored object within the container.

Some biomimetic applications use more than one pattern of the same biological systems. For example, the water vapor harvesting device integrates few patterns that appear at the Namib Desert beetle including the hydrophilic bumps (repeated protrusions), the waxy channels between these bumps (repeated channels) and the angle of the beetle's body (Asymmetry).

8.4.4 Findstructure Database

The research database (the observation database) including the 140 biological systems was uploaded to a website, the Findstructure database [193]. Tens new biological systems were added to the website since the termination of the study, as part of an ongoing elaboration process.

Findstructure is a biomimetic database that supports biomimetic design in general and the structural biomimetic design method in particular. This database provides biological functional solutions, already analyzed per nature structure-function patterns. It can be used by designers, engineers, scientists, architects, or biologists, who search for design solutions in nature, and need a source of both knowledge and inspiration to facilitate technological innovations. Based on the structure-function table typology (Table 10.4), each biological system in the database is categorized by the structural patterns and related generic functions. These patterns often abstract the richness and complexity, but nevertheless, they embed some design principles and foster the abstraction stage of the biomimetic design process. Further elaboration on the Findstructure database searching process is introduced in Chap. 10.

8.5 Discussion of Results, Explanations and Implications

Based on numerous analyses of biological systems by the complete viable system model, we identified several structure-function patterns that repeat in biological systems generally and in biomimetic applications particularly. No additional patterns emerged during the analysis of the 140 biological systems in our observations database but the ones presented in Table 10.4. Furthermore, since the analysis completed, all examples we have encountered, could be fitted into the same structure-pattern set.

These structure-function patterns repeat in different scales. For example, repeated protrusions appear at a macro scale (the burrs of the burdock plant), or at the nano scale (epidermal protrusions), with exactly the same function of attaching/detaching. Repeated channels appear at the macro scale (the termite's channels) or at the microscale (wood tubes or membrane cell), and similarly, they have the same function of channeling/regulating. Helical structures appear at the macro scale (horns) and at the microscale (collagen fibers), again, with the same function of protection/absorbing shocks. Though the scale may affect the structural characteristics [160], it is interesting to realize that at the meta-level, the same structure-function patterns appear in different scales.

Some of the structures were defined as engines, exploiting external fields of energy for performing dynamic functions, while others were identified as brakes, blocking external fields of energy that may damage the system, thus exhibiting static functions.

The idea of nature engines is mainly known at the cell level. For example, the plants movement towards the sun (Heliotropism) is a result of motor cells that change their turgor pressure by pumping ions and changing the ions concentrates. Vogel [164] used the concept of nature engines for engines that are based on natural forces like hydration devices or osmotic devices. He mentioned the motile cilia, microscopic cylinders that can actively bend and used for locomotion [194]. We provide examples of nature engines at the macro level as structures that exploit external energy sources. Finally, we generalize the examples into recurring engine patterns; repeated protrusions, repeated tubes and asymmetric structures.

The idea of nature brakes presents the opposite idea of an engine. As engines exploit forces, brakes reject forces. If engines exist in nature, then we might expect to find also brakes. Brake structures resist external loads and inhibit their undesired resulting motion by absorbing, pushing back or changing the direction of these loads. Most of our identified brake structures are already known as structures with good streaming or mechanical properties, but they might have not been grouped before as structures that inhibit motion, or called brakes. Observing a single tube exposes an interesting duality as a tube is an engine structure when placed in the flow orientation and a brake structure when it resists mechanical loads that may cause deformations. Another example of duality is muscles that can be contracted or released during time, demonstrating time asymmetry. When muscles transduce energy, they act like engines and when they absorb energy, they act like brakes. This is another interesting example of duality in nature, when one structure serves both functions, in different points of time. Both duality examples represent an efficient use of space and material resources for executing multiple functions.

These engine and brake structures are related to high level function ontology. As oppose to Gorb [170] that related biological surfaces to survival functions such as feeding, filtering or self-cleaning, we view functions through a technical lens as basic interactions between the biological system and various substances in its surrounding.

This technical view allowed us to abstract some patterns. Thus for example, while the grooved scales of the shark skin are related to drag reduction, gecko's hairy foot is related to adhering, epidermal protrusions are related to self-cleaning and hairs are related to filtering, we abstract all these above mentioned structures to a general pattern of repeated protrusions, whether these protrusions are scales, hairs or epidermal protrusions. Respectively, we abstracted all these aforementioned functions to a general function of attachment or its opposite detachment, whether the function object is flow vortices that are attached to the shark skin, the gecko's foot that is detached from the surface, dirt particles that are detached from the leaf or food particles that are attached to the filter hairs. Whatever is the purpose of attachment (filtering, self-cleaning etc.) and whatever is the object being attached or detached, we recognize the same structure of repeated protrusions.

While all the structures of the structure-function patterns were identified as engines or brakes, in some cases they served in addition as the working unit of the system, demonstrating a unification of system parts. One example for this unification is the click beetle jumping mechanism (Fig. 8.5). The muscles are a time asymmetry structure as they demonstrate different appearances at different times, contracted or not, and they serve as both the engine and the working unit of the system. We distinguished between environmental and non-environmental control unit [176]. In some cases the control component is part of the organism itself, mainly the organism nerves system. In other cases the system responds to environmental factors, such as humid levels, so the control is done by environmental conditions. In the brake model, control is done by the presence of the harmful factor that may cause the harmful function. In that case, the braking action is taken place.

The list of structure-function patterns (Table 10.4) summarizes our results and provides a high level abstraction of structure-function relations in biomimetic sources in general and in biomimetic applications in particular. As all patterns, they allow designers to disconnect from the details and use the abstracted design principles in bio-inspired design processes. Patterns always come at the expense of richness and complexity, but on the other hand they extract some design principles and foster the abstraction stage of the biomimetic design processes. Structure-function patterns (Table 10.4) may be used by biomimetic designers in several ways:

(1) List of extracted bio-inspiring design principles regarding the relations of structure and functions. For example, a designer who is familiar with the relation of repeated protrusions and attachment/detachment structures may use this knowledge to design concepts of innovative shoes without imitating specific biological organism.
(2) Abstraction of biological systems—Yen et al. [20] reported on difficulties to clearly identify the functions of the biological system in an abstracted way that could sustain the transfer of design principles. The structure-function patterns can assist designers to understand the functional mechanism, by identifying structures at biological systems and related functions. Biological systems usually integrate several structure-function patterns that create complex functional mechanisms. An example for intertwined patterns in the same system is the blood artery that has a tubular structure that channels the blood whereas its walls are made of elastic helical collagen that protects it from the blood pressure [185], [195].

(3) "Front-end" index to thorough studies—The structure-function patterns may serve as a "front-end" index to thorough studies such the one performed by Gorb [170]. Thorough studies add knowledge about the mechanism behind these structure-function patterns. For example, it was found that enlargement of the contact area in attachment mechanisms is achieved by splitting the contact into sub contacts according to contact theory [196]. In light of this explanation, it is clear how the repeated protrusions enlarge the contact area.

Other important results of this part of the study are:

(1) Abstraction tool—The complete viable system model has a value as an abstraction tool for representing biological solutions. It is based on a technical view that is inherent and natural to engineers and explains the role of the structures in a complete system model and their relation to the system function. The complete viable system model is actually based on a SBF view (Sect. 3.2.2.1), as it provides behavioral explanation for how the structures accomplish the function.
(2) "Findstructure" biomimetic database—The observations database itself is a biomimetic database with biological systems. Designers may search it by functions, by structures or by the structure-function patterns. It can be accessed by biomimetic designers as a searching tool, not only for the ones who use the structural biomimetic design method.

Chapter 9
Sustainability Patterns

In this chapter we offer the concept of 'Ideality', presented in Sect. 6.1.3.4, as a framework that can guide the search for sustainability patterns in nature [197]. We explain the logic of using this framework and demonstrate how it enriches the current knowledge base of the life principles with new insights. Ideality analysis of selected examples of biological forms and structures is presented and a list of ideality strategies that repeat in nature is summarized and compared to the life principles knowledge base.

9.1 The Analysis Process

The analysis process relies on TRIZ ideality law as a guideline. We assume that if technological systems evolve towards more ideal situation, biological systems had much more time to evolve towards ideal systems in their environments.

9.1.1 The Analysis Rationale: On the Relations of Ideality and Sustainability

Sustainability and ideality are basic notions in design. While ideal systems had always been aspired to, sustainable systems are relatively new demand of design. In this section we explore the similarity and differences between these two basic notions and suggest that there is a strong correlation between ideality and sustainability. Based on this correlation we suggest ideality as a framework of reference for analyzing sustainable aspects of biological systems, and transferring the results of this analysis back to technology to assist in designing sustainable products.

Y.H. Cohen and Y. Reich, *Biomimetic Design Method for Innovation and Sustainability*, DOI 10.1007/978-3-319-33997-9_9

Sustainability is defined as the ability to be used without being completely used up or destroyed, or the ability to last or continue for a long time (Mirriam, Webster Dictionary). Charter and Tischner [37] defined sustainable solutions as: "products, services, hybrids or system changes that minimize negative and maximize positive sustainability impacts—economic, environmental, social and ethical—throughout and beyond the life-cycle of existing products or solutions, while fulfilling acceptable societal demands/needs." Their definition related the notion of sustainability to the concept of ideality. The TRIZ definition of Ideality is the qualitative ratio of all systems useful functions to harmful functions or simply the ratio of systems benefits to costs [23]. By maximizing positive effects and minimizing negative effects the product is more sustainable according to Charter and Tischner [37] or more ideal according to the TRIZ definition for ideality. Hill [99] expressed a similar idea when he defined the effectiveness of biological structures as a cost-benefit relation between maximization of the 'survival functions' (including sub-functions such as reproduction, feeding, defense, movement etc.) and minimization of costs (energy use and biomass).

The relation of Ideality to sustainability is clear. Assuming that resources are not unlimited, a relevant strategy to sustain is to "achieve more with less" meaning achieving more benefits with less resources. Therefore, sustainability may be achieved by ideal or effective systems that provide more benefits at low costs.

The relation between Sustainability and ideality has already been a base for the development of practical eco-guidelines for product innovation and sustainability [140, 198]. Here we offer to use this relation for biomimetic design. The ideality definition is applicable for biological systems that demonstrate various strategies to increase their benefits on one hand and to reduce their costs on the other hand. Due to a competition on resources and the fact that some of the used resources are not renewed at the required rate, biological systems must demonstrate ideal and efficient architecture, structures and processes in order to sustain. While searching for general sustainability strategies or life principles is difficult or not always clear, searching for ideality strategies is well directed by a simple principle of increasing system benefits and reducing system costs.

9.1.2 The Analysis Stages

We analyzed biological systems using the ideality definition as a framework to guide our search. We analyzed by the ideality framework several biological systems from our observations database, few of each structural pattern (Sect. 8.4). We chose to focus on these systems as they represent a sample of generic structures that repeat in nature in different systems and scales, demonstrating nature efficient solutions to stringent limitations on the design space. As these structures are generic, they might represent generic ideality strategies as well.

Table 9.1 Technological ideality strategies, extracted from the TRIZ literature [145, 199]

	Ideality strategy
Increase benefits	Adding functions to the existing working parts
	Improve the performance of some functions
Reduce costs	Exclude auxiliary functions that support the main function but may be removed without affecting the performance of the main function
	Combine subsystems of several functions into a single system
	Transferring some functions to a super system
	Utilizing internal and external resources that already exist and are available
	Use of physical, chemical, geometrical and other effects as resources
	Improving the conductivity of energy through the system (provide easier access)
	Synchronization system parameters to prevent waste

For each biological system we used a two stage analysis. First, we searched for general intuitive strategies that increase benefits or reduce costs. We asked in what ways the benefits of this system are increased and the costs are reduced. Then we used the reference of ideality strategies that repeat in technological systems (Table 9.1) and tried to identify similar strategies in nature. Selected examples of ideality analysis are presented in Table 9.2 per various structural patterns.

9.2 Results—Nature Ideality Strategies

Nature ideality strategies and design principles that repeat in our sample of structural patterns are presented with examples in Table 10.10, as part of the structural biomimetic design method manual.

9.3 Results—Ideality Tool for Sustainable and Biomimetic Design

9.3.1 Ideality as a Tool for Sustainable Design

The ideality strategies (Table 10.10) are sustainability patterns extracted from nature. They are suggested as sustainability tool adjusted to the early concept design stage, as it does not require detailed information about material and energy flows. A designer may use the ideality strategies as a checklist and implement them at the design concept as design principles that foster sustainability. Thus, the ideality strategies are applicable for general design processes, not only for biomimetic design processes.

Table 9.2 Selected examples of ideality analyses per various structural patterns

	Structural patterns	System example	Ideality analysis (Each ideality principle has a number as presented in Table 10.10. Therefore the numbers in this tables are not sequential)
1	Repeated protrusions	Lotus leaf [53]	*Increase benefits* 1. Multifunctional Design—The epidermal protrusions provide both dirt removal and protection against bacteria [200] 2. Intensified Interaction—Stronger effect of dirt removal is achieved by the repetition of protrusions *Reduce costs* 3. Reduction of external disturbance—Protrusions remove dirt that can harm the photosynthesis process. Removal of harmful bacteria 5. Using available effects and gradients as energy resources—Using adhesion and gravitation gradients for the dirt removal function 6. Adjustment of structure to function—Epidermal protrusions create superhydrophobic structure, adjusted to dirt removal 7. Transferring some functions to the supersystem—The water droplets, part of the supersystem, serve as the working unit that actually removes the dirt particle 8. Synchronizing system parameters to prevent waste—The removal of dirt is synchronized with the appearance of water droplets to prevent waste of light energy
2	Repeated channels/tubes	Termite mound [22]	*Increase benefits* 1. Multifunctional design—Tunnels provide both stability and a path for the air flow 2. Intensified interaction—The repetition of tunnels increase the surface area of air flow, achieving stronger effect of gas exchange regulation *Reduce costs* 3. Reduction of external disturbance—Tubular structure of the tunnels system protect against external harmful loads 5. Usage of physical gradient—Using temperature gradient to support transfer air. Saving energy cost of air movement 6. Adjustment of structure to function—Tunnels network is adjusted for channeling. Their tubular shape is adjusted to absorb mechanical loads

(continued)

Table 9.2 (continued)

	Structural patterns	System example	Ideality analysis (Each ideality principle has a number as presented in Table 10.10. Therefore the numbers in this tables are not sequential)
			7. Transferring some functions to the supersystem—Using local ground to build the mound. Fungi that live inside the mound provide food and get a shelter (symbiosis) 8. Synchronizing system parameters to prevent waste—Air flow is synchronized with the surrounding temperature, preventing over heating or cooling 9. Improving the conductivity of energy through the system by the network of tunnels that provide easier access for the air flow
3	Asymmetric structures	Pine cone scales [190]	*Increase benefits* 1. Multifunctional design—Pine cone scales contain, protect, and release the seeds 2. Intensify the interaction—Repetition of scales intensifies contain and release of seeds *Reduce costs* 3. Reduction of external disturbance—Protect the seeds within container (scales) 5. Usage of physical effect—The pine cone scales are opened due to hygroscopic expansion gradient between the inner and outer scale layers 6. Adjustment of structure to function—Scales structure is adjusted to the opening and closing movement 7. Transferring some functions to the supersystem—The water, part of the supersystem, serves as the working unit that actually swells the cells and open the scale 8. Synchronizing system parameters to prevent waste—Scales opening and closing are synchronized with the humid levels, thus preventing waste of seeds released in time not appropriate for germination
4	Repeated layers	Abalone shell [179]	*Increase benefits* 1. Multifunctional design—The shell contains and protects the animal. Provide both mechanical and thermal protection 2. Intensified interaction—More protection is achieved by the repetition of layers as the loosely layers absorbs the loads, reduces the level of loads and prevent failures or cracks *Reduce costs* 3. Reduction of external disturbance—Layers reduce loads that may cause cracks, deformation and more 6. Adjustment of structure to function—The hard layers slide instead of breaking and the protein stretches to absorb the energy of the pressure 9. Improving the conductivity of energy through the system by the elasticity of layers

(continued)

Table 9.2 (continued)

	Structural patterns	System example	Ideality analysis (Each ideality principle has a number as presented in Table 10.10. Therefore the numbers in this tables are not sequential)
5	Container	A sphere [162]	*Increase benefits* 1. Multifunctional design—Spheres create a cavity that both contain an object and protect against external loads *Reduce costs* 4. Reduction of surface area when it is harmful—Sphere is known as the shape that has the least possible surface area for a given volume [157] resulting in a reduced material & weight and less exposure to predators 6. Adjustment of structure to function—A sphere has the ability to absorb and dissipate loads due to its curvilinear form [162] 10. Give up redundant parts—The reduction of surface area give up redundant material
6	Streamlined shapes/structures	Penguin contour/fish schools/river basins	*Increase benefits* 1. Multifunctional design—The penguin contour creates a cavity that both contains the internal organs and reduces turbulences *Reduce costs* 3. Reduction of external disturbances—The penguin body contour and fish school structure minimizes resistance to motion through water 5. Usage of physical effects as gradients—Fish schools creates positive constructive hydrodynamic between the wakes of neighboring fish and use it as a source of energy for propulsion [201] 6. Adjustment of structure to function—Most of the streamlined shapes are consisted of curvilinear forms that have the ability to absorb and dissipate loads [162] 9. Improving the conductivity of energy through the system—Easier flow in river basins (network structure), corresponding to Bejan's constructal theory [8]

(continued)

Table 9.2 (continued)

	Structural patterns	System example	Ideality analysis (Each ideality principle has a number as presented in Table 10.10. Therefore the numbers in this tables are not sequential)
7	Cylinder/tube	Tree roots	*Increase benefits* 1. Multifunctional design—Trees roots provide stability, easier penetration and channeling of water and nutrients *Reduce costs* 3. Reduction of disturbance—Tube structure reduce loads by providing stability against bending and buckling and resists bending in all direction [161] 6. Adjustment of structure to function—The tubular structure of the tree's roots is adjusted to the function of penetrating into the soil 9. Improving the conductivity of energy through the system—Easier flow in the roots network structures, corresponding to Bejan's constructal theory [8] 10. Give up redundant parts—Better strength is achieved by giving up redundant inner solid filling of a tube and saving material. In this case, the respective distance of compression and stretch is larger in the hollow tube and it is more difficult to bend it [164]

For example, the challenge of sustainable cities may be observed under the ideality lens and elaborated with the ideality sustainability tool. In that case, the ideality tool extends the scope of sustainability and may suggest operative design ideas, not necessarily related to traditional definitions of sustainable cities, such as:

- Provide more benefits to their citizens and not only focus on minimization of resources and wastes. Such physical, social or economic benefits may be clean air, shade, light, community support and abundance of work possibilities.
- Save the cost of potential disturbances such as car accidents, noise, heat, congestion.
- Use available resources such as sun and wind energy, clean water, and even the knowledge resources of its collection of citizens.
- Synchronize city parameters such as the opening hours of various institutes to prevent traffic congestions.
- Give up redundant areas that provide no benefits but require costs, such as contaminated lands.
- Improve the conductivity through the city by providing easier access to energy and information resources, as well as mobility services.

This example shows how the ideality bioinspired tool may assist urban planners and policy makers to define design strategies for sustainable/ideal cities. Product designers may also use the ideality tool for sustainable design. A detailed example is the following case study of multiphase bicycle.

9.3.1.1 Sustainable Design Case Study: An Ideal Multiphase Bicycle

This case study was a class experiment during a course in engineering design in the third year of mechanical engineering degree. Twenty students got an explanation about the ideality tool from an industrial designer, highly experienced in design processes and familiar with the ideality tool. Later, students were asked to use the ideality tool during a design process.

The Design Challenge: The Multiphase Bicycle

Current bicycle products are adjusted to various mental and physical development stages during childhood. As a result, parents are required to purchase several different bicycles from the age of two till the child acquires the ability to ride independently and pay the extra costs of replacing bicycles frequently. There is a need for a multiphase bicycle that can be adjusted to a longer period during childhood. Therefore, the design task was to offer a design concept for a multiphase children bicycle, by using the ideality tool. The suggestions for the multiphase bicycle design concept, presented in Table 9.3, were developed during the class experiment and elaborated later by the industrial designer who was in charge of this experimental session, as well as by the authors.

Table 9.3 Design concept suggestions for an ideal multiphase bicycle

Increase benefits	
More functions	**Multifunctional design** 1. Changeable bicycle (size, shape, body elements)—On one extreme, parts as the auxiliary handle or wheels may be removed or replaced (like slough in biological systems). On the other extreme, system parameters may be changed during time, such as the size and height of the bicycle handlebars and seat, the distance between wheels, or the size of the wheels, (like growth of biological systems) [How: May be achieved by actuators, folding mechanisms, or inflatable chassis or wheels.] 2. Amusing bicycle—During infancy, adding function of amusements to a regular ride may encourage infants to practice. Such amusements may be achieved by adding music or art activity. The board wheel surface, for example, may be used as a board to sketch a map or impressions during the ride breaks. Later on it may be removed 3. Feedback bicycle—providing feedback to rider relating to ride parameters: stability, velocity, distance. Different feedback parameters are required in different ages
Stronger effect	1. Stronger grasping effect (for early stages) by repeating tire protrusions and grooves, to achieve larger surface contact. Protrusions patterns may be changed later when changing the wheels 2. Stronger grasping effect (for early stages) by repeating wheels that can then be replaced later by a single wheel to reduce friction. Wheel may be assembled one beside the other to get a larger surface area (or even as two wheels), and later on removed
Reduce costs	
Defensive strategy	**Reduction of disturbances** 1. Reduce possible damages to the bicycle, so it will last longer period during the use: Prevent wheel punctures [How: by airless tires made of fully structured material] Defend the chassis against corrosion and other weather damages [How: through coatings technologies] 2. Reduce possible injuries, mainly during the training period: Prevent contact of clothing or legs with shrubs
Opportunistic strategy	**Usage of physical, chemical, geometrical and other effects and gradients as energy resources** 1. Using the cycling energy for transportation/light in older ages when riding distances are significant [How: via Dynamo] 2. Using heat energy in early stages
	Adjustment of structure to function Reduce the number of elements that are needed to handle growing to allow only their replacement. Locate these elements in convenient locations for replacement
	Transferring some functions to the super-system In early stages, transferring the steering function to the accompanied parent [How: by steering leading stick]. Later on, bringing back steering functionality to the bicycle handlebars

(continued)

Table 9.3 (continued)

Prevent waste	**Synchronize system parameters to prevent waste** 1. Synchronizing the bicycles characteristics such as size, height and auxiliary devices to the child age, prevents waste of money 2. Synchronizing item changes with other items to allow a change in the system without modifications 3. Synchronizing bicycles feedback with riding performance—Emotional and physical self-feedback system provides the rider a feedback on his stability and balance [How: vocal feedback system]. This feedback system prevents an emotional waste of frustration, physical waste of injuries and time waste by a shorter learning curve
	Improve the conductivity of energy through the system to provide easier access and prevent waste of energy One difficulty of children is to stop the pedals in the right position to launch the ride. Pedals positioning system locates the pedals in the right position for launching rides. The child just needs to step on the pedal and does not need to locate it in the right position. This feature may facilitate the ride in young ages. It can be abolished in older ages when eye-body coordination is developed
	Give up redundant parts 1. Unnecessary parts such as auxiliary handling system or auxiliary wheels are removed when the child is getting older and does not need them 2. Service of bicycle rental—rent various bicycle in different ages 3. Improve circulation of 2nd hand bicycles 4. Cradle to cradle production, use and recycling strategy

The industrial designer reflected that using the ideality tool yielded innovative product features such as pedals positioning system or self-feedback system that would be doubtfully defined without the ideality tool. Thus, the tool served as innovation leverage, beyond designers thinking fixations. Using the ideality tool yielded also sustainable and resilient features. The proposed bicycle concept extended product lifecycle due to the multiphase modular design and the damages prevention.

The industrial designer reflected also on the design process. The tool supported mainly the product specifications stage, a critical stage that affects the visible value of the product. In this stage, the tool served as a compass for ideal design, forcing designers to focus on "What" makes a product an ideal one and not "How" to make it. A natural tendency to move directly to the "How" stage is prevented. In Table 9.3, content related to the "How" stage is mentioned in brackets while most of the content refers to the "What" stage. The "How" related ideas may be elaborated later in the design process, after the final specifications of the product is agreed upon.

The ideality approach for sustainable design, has already been developed into a design tool, the ideality "WHAT" model for sustainable product brief [202]. The tool offers a checklist based on the intersection of ideality strategies and product brief categories. This intersection encourages designers to examine how each ideality strategy can be realized in each aspect of a product design brief.

9.3.2 Ideality as a Tool for Biomimetic Design

In relation to biomimetic design, a designer may analyze the ideality aspects of a specific biological role model by the ideality framework (qualitative ratio of benefits to costs) and by the assistance of Table 10.10 as a reference. Then, the designer should repeat this ideality analysis process for the related biomimetic design concept. The comparison between these two analyses reassures that the sustainability strategies of the biological model are kept and transferred to technology. In a metaphoric way we use this comparison to reassure we do not lose some degrees of ideality during the transfer from biology to technology, as sustainability is not guaranteed. A designer may imitate an innovative design concept from nature but perform it using harmful materials and pollutant manufacturing processes. Such a comparison between the ideality analysis of the biological and biomimetic models is a measure to evaluate the sustainability transfer.

Besides the current list of ideality strategies, the ideality framework itself may guide the search for more ideality strategies in nature, enriching Table 10.10 with new insights. We may analyze different biological systems and ask how do these systems increase their benefits and how do they reduce their costs. In this process we may identify new ideality strategies, not identified before. Whereas the life principles are a closed core of knowledge, as a designer may only use them but not add new principles, the ideality framework is an open core of knowledge.

Nature ideality strategies are integrated within the structural biomimetic design method to reassure the transfer of sustainability strategies from biology to technology or other domains of applications, during the biomimetic design process. Detailed explanation for this integration may be seen in Chap. 10. Examples for this integration are seen in Chap. 11.

9.4 Discussion of Results, Explanations and Implications

9.4.1 The Ideality Strategies Characteristics

We used ideality as a framework for sustainability analysis and provide a list of ideality strategies and design principle patterns that repeat in nature (Table 10.10). Increasing ideality by reducing costs is more prevalent in nature (8 design principles) compared to increasing ideality by increasing benefits (2 design principles).

The cases presented in Table 9.2 demonstrate more strategies to reduce costs but in all cases there is at least one strategy to increase benefits. Some ideality design principles are more frequent, such as ‘Multifunctional design’ and ‘Adjustment of structure to function’ that appear in all cases and considered to be basic design principles in nature.

Table 9.4 Comparison between sustainability strategies defined by the complete viable system to sustainability strategies defined by the ideality model

	The complete viable system model: sustainability strategies	The ideality model: sustainability strategies
1	Utilizing environmental energy sources	Usage of physical, chemical, geometrical and other effects and gradients as energy resources
2	Utilizing environmental elements as essential systems parts	Transferring some functions to the supersystem. Using supersystem material resources
3	Unification of system parts	Give up redundant parts

It is also interesting to realize that most nature ideality strategies (Table 10.10) have matching strategies in technology (Table 9.1). Using ideality strategies in technology as a reference was valuable and illuminated similar strategies in nature.

Ideality strategies correspond to the idea of engines and brakes in nature. Two general ideality strategies of opportunistic and defensive strategies present a similar idea of engines and brakes respectively. Nature engines use physical, chemical, geometrical and other effects and gradients as energy resources (Ideality strategy #5). Nature brakes are structures that reduce disturbances such as friction, loads, turbulence and more (Ideality strategy #3). Ideality strategies correspond also to sustainability strategies defined by the complete viable system (Sect. 8.3). The similarity is presented in Table 9.4.

9.4.2 *Ideality as an Evolution Law*

The law of ideality is one of the laws of technical evolution, part of the TRIZ knowledge base. It states that every technological system tends to develop towards increasing the degree of ideality, trying to achieve the ideal final result, a hypothetical state of having all benefits at zero cost. Accordingly, natural systems may be directed by the same law of ideality, increasing their benefits and reducing their costs during evolution. This principle may enrich our general understanding of nature structures formation in general and patterns formation in particular.

9.4.3 *Ideality Strategies Versus Life Principles*

Comparing ideality strategies in nature with the life principles (Fig. 1.4) reveals the similarities and differences of these two approaches. Table 9.5 presents the similarities and differences between the life principles and nature ideality strategies.

Table 9.5 shows that the ideality framework partially overlaps the life principles and in some cases completes and enriches it. For example, 'Multifunctional design' and 'Adjustment of structure to function' appear in both lists with similar wording.

Table 9.5 Comparison of life principles and nature ideality strategies

		Life principle	Nature ideality strategies/principles
1	Related strategies	Use multifunctional design	Multifunctional design
		Fits form to function	Adjustment of structure to function
		Use readily available materials and energy	Usage of physical, chemical, geometrical and other effects and gradients as energy resources (Saving energy costs)
		Cultivate cooperative relationship	Transferring some functions to the supersystem
		Use feedback loops	Synchronizing system parameters
		Build selectively with a small subset of elements	Give up redundant parts
		Embody resilience through variation, redundancy, and decentralization	Defensive strategy—prevent disturbances: 1. Reduction of external disturbance such as friction, loads, turbulence and more by structures 2. Decrease of surface area when it has harmful effects such as extended exposure to predators
2	Appears in nature ideality strategies and not in life principles		Intensify the interaction with the environment to achieve stronger effect of one function by: (1) repetition of elements; (2) increased surface area
3	Appears in life principles and not in nature ideality strategies	✓ Replicate strategies that work ✓ Integrate the unexpected ✓ Reshuffle information ✓ Use low energy process ✓ Recycle all materials ✓ Maintain integrity through self-renewal ✓ Incorporate Diversity ✓ Combine modular and nested components ✓ Self-organize ✓ Build up from the bottom up ✓ Combine modular and nested components ✓ Leverage cyclic processes ✓ Break down products into benign constituents ✓ Do chemistry in water	

In other cases similar principles appear in both lists but the life principles are more general while the ideality strategies are more operative and descriptive. For example, the life principle of "Use readily available materials and energy" may not be operative enough as it does not provide the answer where these available energy and material resources may be found. In contrast, the related ideality principle of using physical, chemical, geometrical and other effects and gradients as energy resources provide the answer where to find available resources. Another example is the life principle "Use feedback loops" that may not be operative enough as it does not provide the answer how to use the feedback loops. In contrast, the related ideality principle of "Synchronizing system parameters" suggests using feedback loops by synchronizing system parameters with environmental parameters.

In some cases, the ideality framework enriches and completes the life principles with extended design principles for the same subject. For example, while reacting to disturbances appears in both frameworks, the life principle suggests managing disturbances by resilience through variation, redundancy, and decentralization. The ideality principle suggests reducing external disturbances such as friction, loads, turbulence, and more by adjusting structures and by decreasing surface area when it has harmful effect. In other case, through the lens of ideality, we revealed a new sustainability design principle in nature: Intensify the interaction with the environment to achieve stronger effect by repetition of elements and/or by increased surface area. In addition to achieving more functions by each structure (multi-functional design), there is also a tendency to optimize the effect of each function.

Life principles are based on a holistic approach viewing the organism as part of its ecosystem. The ideality framework is based on a technical and functional view. For example, the life principle "Cultivate cooperative relationship" is based on a holistic approach of Win-Win relationship. The related ideality principle of "Transferring some functions to the supersystems" mainly focuses on the benefits to the system though the supersystem element may be also an organism that gains a function through a symbiosis relationship.

Life principle strategies that do not appear in nature ideality strategies mainly refer to the growth and manufacturing processes such as recycling, self-organization and water based chemistry. One way to understand this missing core of principles is the fact that we observed current functioning of structures and did not observe how did they form or end their life. However, the framework of ideality may identify these principles if we extend the scope of analysis to include also processes. For example, recycle all materials is relevant to the ideality strategy of "Prevent waste for better usage of resources". Generally, life principles include more strategies and design principles, but the ideality strategies may be extended in future research as the searching framework is simple and clear. On the other hand, it is not clear how to search for more life principles.

In summary, the ideality framework is a simple framework of reference and a way of thinking that may be used for the process of identifying sustainability strategies and principles in nature. The ideality framework partially overlaps the life principles and in some cases completes and enriches it.

9.4.4 Ideality as a DFE Tool: A New Approach for Innovative Design for the Environment

Design for the Environment (DFE) is a comprehensive term for various methodologies and tools to integrate environmental factors into products, early before the production stage, such as Design for Recycling, Design for Disassembly, Design for Energy Efficiency and more. In each one, a different product environmental factor is in focus and treated as a design objective, rather than a constraint [203].

In relation to DFE design processes, a designer may use the ideality strategies as a checklist tool and implement them during the conceptual design stage to foster sustainability. But, rather than focusing on one aspect of DFE, the ideality tool provides a comprehensive approach that offer many aspects gathered together under the ideality rational. Some ideality strategies are clearly related to current DFE techniques. For example, ideality defensive strategy to prevent harmful effect is related to Design for hazardous material minimization. Ideality opportunistic strategy is related to design for efficiency, and ideality strategy of preventing waste is related to design for reuse, recycling or disassembly. The ideality strategies that are not related to current DFE techniques may suggest new DFE directions or techniques.

9.4.5 Summary

Table 9.6 summaries the three bases of the structural biomimetic design method:

- Functional patterns
- Structure-Function Patterns
- Sustainability patterns

Table 9.6 The structural biomimetic design method: knowledge structure summary

	Base 1: functional patterns	Base 2: structure-function patterns	Base 3: sustainability patterns
Theories and knowledge bases that support the development of this base	General system theory, TRIZ, functional modeling, function ontologies, Su-field model	General system theory, TRIZ, patterns, patterns based design methods, structure-function patterns, complete viable system model	The TRIZ ideality definition, the law of ideality, the relation of ideality and sustainability
analysis process	For each observation in our database: 1. Identify the main function by the Su-Field	For each observation in our database: 1. Analyze per the complete viable system model 2. Pattern search: Search for structures that repeat in a	For selected observations in our database, few of each structural pattern:

(continued)

Table 9.6 (continued)

		Base 1: functional patterns	Base 2: structure-function patterns	Base 3: sustainability patterns
		functional definition components: the working unit, the target object and the field of energy that enables their interaction 2. Pattern search: Search for functions that repeat in a large number of cases and classify them into high level classes	large number and classify them into high level classes 3. Pattern search: search for repeated relations between these structural patterns and the functional patterns identified in base 1 4. Identify structure-function patterns	1. Analyze per the ideality framework without a reference 2. Analyze per the ideality framework with a reference of ideality strategies that repeat in technological systems (Table 9.1) 3. Pattern search: identify repeated ideality strategies and design principles
Results	Tools	1. Su-Field function ontology (Table 7.3) 2. A method to define functions of biological systems (Table 7.1)	1. The Patterns Table (Table 10.4) 2. Analysis and abstraction model for biological systems based on the complete viable system model (Table 8.2)	1. Ideality patterns Table—list of ideality strategies that repeat in nature (Table 10.10) 2. A framework to search for more sustainability patterns in nature, in any biological system (Fig. 6.4)
	Knowledge	3. Enrich function ontologies literature with the functions derived from biological systems	1. Viewing structure-function patterns relations as part of a complete system model 2. Discrimination between engines and brakes 3. Sustainability strategies (Sect. 8.3)	1. Enrich the life principle knowledge base with new strategies and operative design principles 2. Better understanding of the relations of ideality and sustainability

For each base, we present the theories and knowledge bases that support the development of this base, the analysis process, the resulted tools and the new gained knowledge.

Chapter 10
The Structural Biomimetic Design Method Manual: Process (Flow Charts), Tools, Templates and Guidelines

In Chaps. 7–9 we introduced the three knowledge bases of the structural biomimetic design method. The next step according to our research model (Fig. 5.3) is to integrate these knowledge bases as part of a design algorithm that guides the designer step by step through the biomimetic design process (Module 4). In this section we present the manual of the structural biomimetic design method and explain how to use the three knowledge bases as part of a systematic biomimetic design method. The manual includes:

Process flow charts:

1. From a problem to biology design process
2. From biology to an application design process

Tools:

1. Function-means tree—Design space analysis
2. The patterns table
3. System parts analysis
4. System model analysis: the complete viable system model
5. Ideality analysis
    - Ideality framework
    - Ideality patterns table
6. Transfer platform—Analogy comparison components

10.1 Design Process Flow Charts

Two flow charts are presented. Table 10.1 describes the design process from a problem to biology. Table 10.2 describes the design process from biology to an application.

Y.H. Cohen and Y. Reich, *Biomimetic Design Method for Innovation and Sustainability*, DOI 10.1007/978-3-319-33997-9_10

Table 10.1 The structural biomimetic design flow: from a problem to biology

General biomimetic design process	The structural biomimetic design method	Tools, templates and guidelines
Problem definition: Identify the design challenge that is required to be solved	**1.** Identify Design Challenge	Describe the design challenge in your words
	2. Identify main function of design challenge	
	3. Analyze the design space and identify possible design paths	**Function- Means tree-** Create a function-mean tree for the design challenge
Bridging to Biology Biomimetic problem definition formulation	**4.** Formulate the design challenge by generic functions and identify related structures	**The Patterns table:** Formulate the design paths by one or more generic functions and identify [illegible]
Biological system: Search a biological model demonstrating an analogical solution	**5.** Search for a biological system that is identified with this structure-function pattern	1. "FindStructure" (Observations database) 2. AskNature
Abstraction Biological solution abstraction	**6.** Identify system parts	System parts analysis (Subsystem, Supersystem)
	7. Identify structures & functions	The patterns table
	8. System model according to type of structure: Engine or brake.	The Complete Viable system model
	9. System behavior	How do the structures operate to achieve the function?
	10. Ideality analysis	Ideality tool: Ideality framework & Ideality patterns table
Transfer Transfer the solution to the biomimetic application	**11.** Describe biomimetic design concept. Repeat stages 6-10 for the biomimetic concept	
	12. Compare biomimetic concept to biological solution	Analogy comparison tables

Table 10.2 The structural biomimetic design flow: from biology to an application

General biomimetic design process	The structural biomimetic design method (my method)	Tools, templates and guidelines
Biological system: Encounter a biological system	1. Encounter a biological system	A biological system with a unique characteristic and a potential to innovate
Bridging to technology Identify analogical application	2. Identify possible analogical applications. Formulate the main function to be observed in relation to these applications.	
Abstraction Biological solution abstraction	3. Identify system parts	System parts analysis (Subsystem, Supersystem)
	4. Identify structures & functions	The patterns table
	5. System model according to type of structure: Engine or brake.	The Complete Viable system model
	6. System behavior	How do the structures operate to achieve the function?
	7. Ideality analysis	Ideality tool: Ideality framework & Ideality patterns table
Transfer Transfer the solution to the biomimetic application	8. Identify an analogical design challenge for a chosen specific application	
	9. Describe the biomimetic design concept. Repeat stages 3-7 for the biomimetic concept	
	10. Compare biomimetic concept to biological solution	Analogy comparison tables

10.2 Tools

10.2.1 Design Path with Function-Means Tree

The function-means tree presents alternative solutions and a preferred design path solution [204, 205]. Function is signed with ⏢ and mean with □ as shown in Fig. 10.1. The tree is developed by moving from functions to means and vice versa. The design path ends with a mean to realize the function. Ways to develop the tree by separation of branches is suggested in Table 10.3. In addition to separation by functions or means, there might be a separation in space and time. While a classic

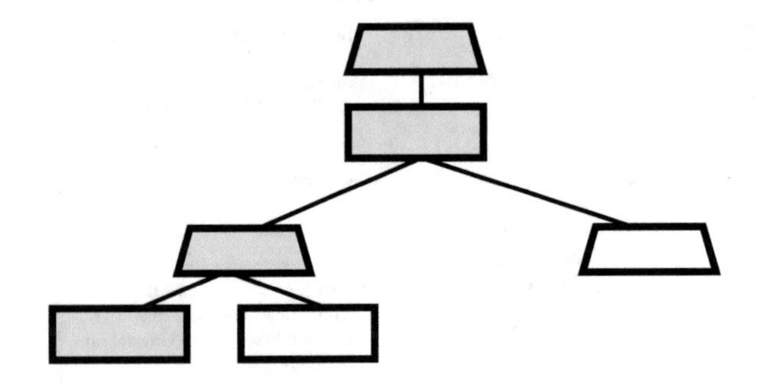

Fig. 10.1 Function/means tree with preferred design path (*shaded*)

Table 10.3 The design path tree: separation types, questions and examples

	Type of separation	Node possible question	Example
1	Separation by functions	Define different functions that are further required by the mean	The cover contains and protects
2	Separation by means	Define different means that achieve the desired function	Slow fall by parachute, slow fall by jet propulsion
3	Separation on time	Define two different states on time	Before/during/after the fall
4	Separation in space	Define two different states in material location in space	There are particles/there are no particles
		Separate between internal and external solutions	Solution with/without external device
			Changing the object/changing the environment

function-means tree is an OR/AND tree, presenting alternative means to realize a function and then functions that each mean further requires, we use a different tree that allows for a more complicated logic between the functions of a mean.

General Remarks: This tool aims to analyze the design space by interpreting the design challenge into possible functions and means. Each branch is a possible design path that directs the search for a biological role model, according to the functions in this path.

10.2.2 The Patterns Table

The patterns table, Table 10.4, is a list of structure-function patterns that repeat in nature in different systems and scales. They were defined by analyzing biological systems with TRIZ models: Su-field analysis and the law of system completeness. They are the core of the structural biomimetic design method and mainly support the abstraction stage.

Table 10.4 The patterns table: list of structure-function patterns

	Structural pattern	Types/private cases	Generic function	Generic functions (second hierarchy)	Generic functions (third hierarchy)
1	Repeated protrusions		Move (Engines)	Attach	Connect, combine, join, adhere, bond, add, increase
				Detach	Remove, subtract, decrease
2	Repeated tubes/channels/tunnels	Without valves		Channel	Lead, guide, direct, flow, stream, transfer
		With Valves		Regulate	Control, modulate, separate, filter
3	Asymmetry	Geometric asymmetry		Change	Change position or location: rotate, spin, turn, move up, move down, move aside, open, close
		Material asymmetry			Change volume or form: blow, blast, cut
		Time asymmetry			
4	Layers (sandwich)		Stop (brakes)	Protect or defend against mechanical or thermal loads	Absorb, push back, resist, isolate, insulate (heat)
5	Intersected layers	Network, cellular, honeycomb			
6	Tube				
7	Helix				
8	Streamlined shapes	Spiral, beak and body contours		Protect or defend against dynamic loads (turbulences)	Stabilize, disperse, deflect, smoothen
9	Container	Sphere, cups		Protect or defend against gravitation/mechanical loads	Contain, store, hold, grasp, trap

10.2.3 Findstructure Database

Findstructure biomimetic database provides biological functional solutions, already analyzed per nature structure-function patterns. It supports the structural biomimetic design method in particular and biomimetic design in general. Searching Findstructure database [193] may be done in multiple ways:

By structures Add one or more of the nine generic structures described in the patterns table (Table 10.4) and find biological systems that have these structures.

It is also possible the search for an integration of some structures in one biological systems by using the 'and' logical operator. For example, "Repeated protrusions and repeated channels" searching results include only biological systems that have these two structures in conjunction, with or without other structures. Using the 'or' operator may be also used in order to extend the search. In order to get more specific results, it is possible to add contextual keywords to the search. Searching by structures is the uniqueness of this database.

By functions Add one or more of the generic functions associated with each one of the nine generic structures described in the patterns table (Table 10.4). Using operators ('and', 'or') is possible in order to narrow or extend the search. Using contextual keywords is also recommended.

By organism Add organism/biological system name with or without contextual keywords. If it appears in the database, you will get a detailed analysis of its functional mechanism in relation to its structures.

Contextual keywords The database supports textual searching so any relevant keywords can be used to search. These keywords may be added to the searching process at any stage or they may be used separately.

Contextual keywords may support the search by providing the right context. They may be related to various factors. Among them are the types of objects involved in the functional interaction, functional specifications or environmental characteristics, as presented in Table 10.5. Each biological system in the database is categorized by one or more of these contextual words. This list of contextual keywords is an open source updated as we add new records to the database.

Searching Results

The Findstructure search results include a list of biological systems adjusted to the searching criteria. You may enter each one of the records and get the following information:

- Biological system name
- Structures—List of identified generic structures, accompanied with their icons (see Fig. 10.2)
- Functions—List of generic functions related to the identified structures
- Function object—the target object of the function
- Context—list of related contextual words based on Table 10.5
- Functional mechanism—full description of the biological functional mechanism described by nature's structure-function patterns
- Biomimetic innovation—link to related biomimetic innovations, if they exist
- Reference—link to general or academic references including papers, books or other sources
- Images—Images of the biological system or functional mechanism illustration.

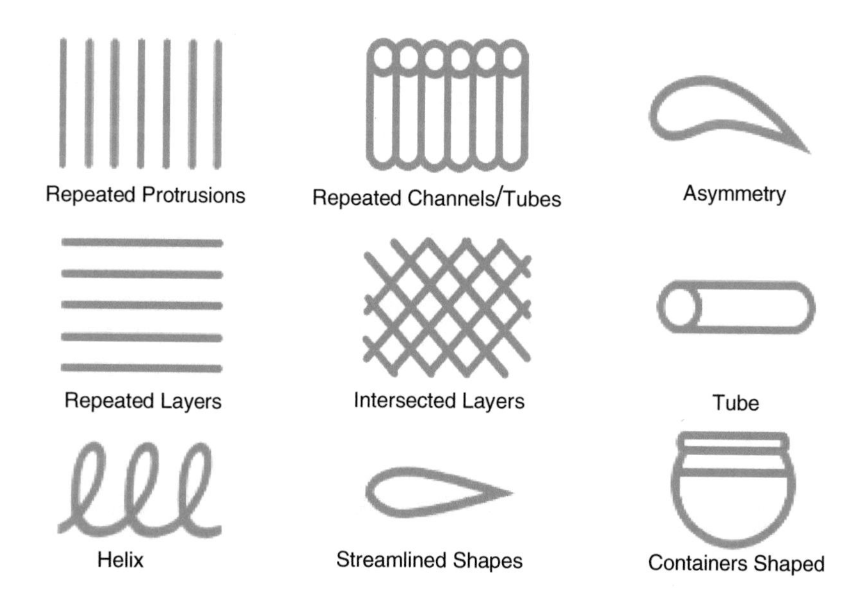

Fig. 10.2 Findstructure icons

Table 10.5 Findstructure contextual keywords

Type of objects	Functional specifications	Environment characteristics
Light	Abrasion resistance	Arid
Material	Anchoring	Cell
Water	Anti-adhesive	Desert
	Anti-fouling	Humid
	Anti-reflective	Hygiene
	Bio-fouling	Lubricated surfaces
	Blood regulation	Marine
	Deformation	Mechanical loads
	Drag reduction	Sand
	Dry adhesion	Smooth surfaces
	Efficient movement	Wet
	Filtering	Wind
	Floating	
	Gas exchange	
	Harvest water	
	Hygroscopic movement	
	Open/close	
	Pain reduction	
	Rain harvesting	

(continued)

Table 10.5 (continued)

Type of objects	Functional specifications	Environment characteristics
	Rotational movement	
	self-cleaning	
	Self-irrigation	
	Spin	
	Temperature regulation	
	Vertical movement	
	Water resistance	

Table 10.6 System parts analysis: definitions, guiding questions and examples

Template	Definitions	Guiding questions	Example (Lotus leaf)
Supersystem Level	Elements that are part of the surrounding environment of the system.	What elements in the surrounding environment of the system are interacting with the system?	Dirt particles, water droplets, sun.
System Level	The name of the system we observe (a biological organism or part of it) according to the system borders we defined.	What are the system borders?	Lotus leaf.
Subsystem Level	Elements that are part of the system defined borders.	What is the system consisted of?	Epidermal protrusions, wax cover.

10.2.4 System Parts Analysis

The abstraction stage of the biological system starts with identifying system parts. The following framework defines two levels to search for system parts, the supersystem level and the subsystem level. The framework presented in Table 10.6 provides both definitions and guiding questions for this system parts analysis.

General Instructions:

1. The purpose of this analysis is to screen for possible subsystem and supersystem elements that may be required for the functional model of the system.
2. System elements should be substances (not forces for example).
3. When possible, prefer to define the system borders as the organism **physical** borders that differentiate the system and the environment. If your system is an

internal organ, the systems borders will be defined as the physical borders of that organ.

4. Add elements that you suspect may be related to the general function being investigated. You might find out later that an element has no relation to the observed function.

10.2.5 The Complete Viable System Model Analysis

The complete viable system model provides an abstraction platform to model and represent the functional mechanism of the biological system in relation to the structure-function patterns (Table 10.4) and their division to engines or brakes. Hence, there is an engine model and a brake model, presented in Tables 10.7 and 10.8, respectively. The bold lines are components of the Su-Field analysis. The dashed lines are components of the law of system completeness.

Table 10.7 The complete viable system model analysis: engine model

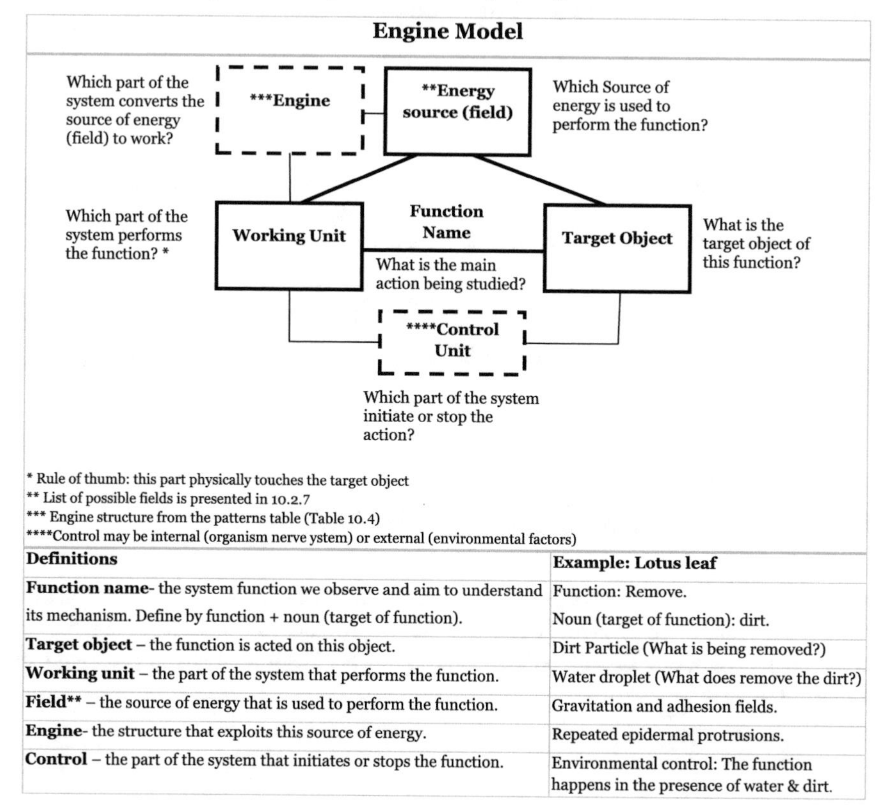

* Rule of thumb: this part physically touches the target object
** List of possible fields is presented in 10.2.7
*** Engine structure from the patterns table (Table 10.4)
****Control may be internal (organism nerve ystem) or external (environmental factors)

Definitions	Example: Lotus leaf
Function name- the system function we observe and aim to understand its mechanism. Define by function + noun (target of function).	Function: Remove. Noun (target of function): dirt.
Target object – the function is acted on this object.	Dirt Particle (What is being removed?)
Working unit – the part of the system that performs the function.	Water droplet (What does remove the dirt?)
Field** – the source of energy that is used to perform the function.	Gravitation and adhesion fields.
Engine- the structure that exploits this source of energy.	Repeated epidermal protrusions.
Control – the part of the system that initiates or stops the function.	Environmental control: The function happens in the presence of water & dirt.

Table 10.8 The complete viable system model analysis: brake model

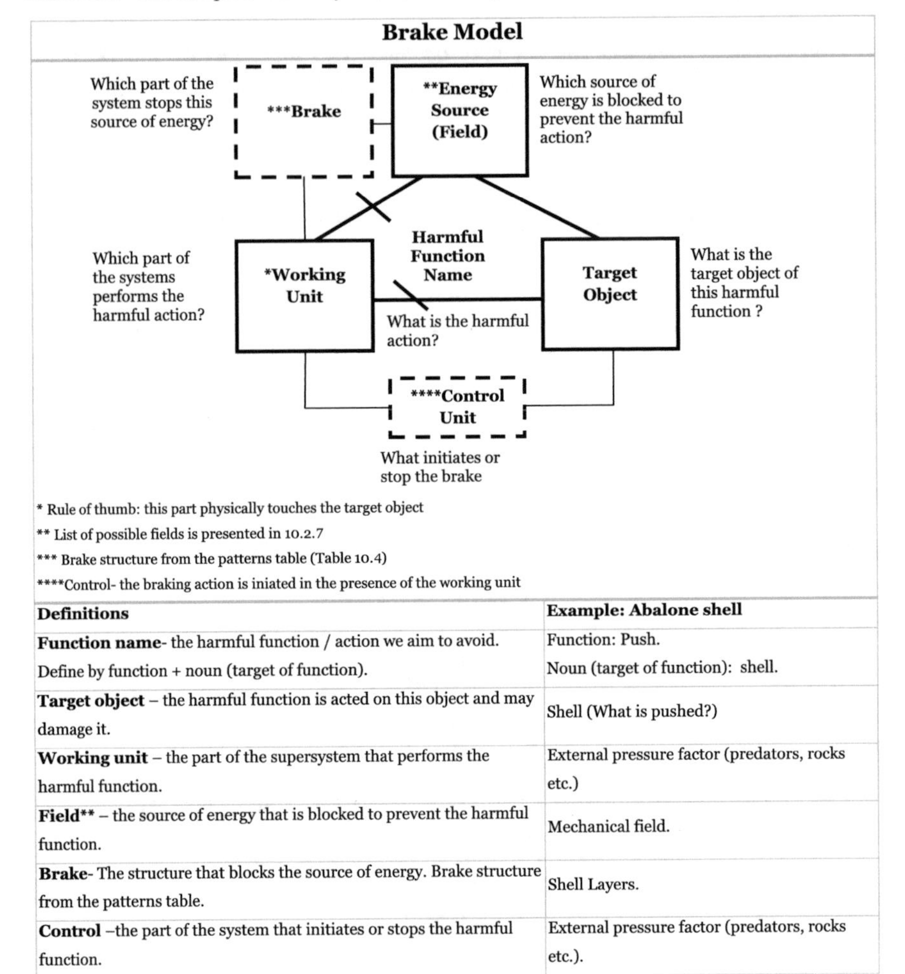

Definitions	**Example: Abalone shell**
Function name- the harmful function / action we aim to avoid. Define by function + noun (target of function).	Function: Push. Noun (target of function): shell.
Target object – the harmful function is acted on this object and may damage it.	Shell (What is pushed?)
Working unit – the part of the supersystem that performs the harmful function.	External pressure factor (predators, rocks etc.)
Field** – the source of energy that is blocked to prevent the harmful function.	Mechanical field.
Brake- The structure that blocks the source of energy. Brake structure from the patterns table.	Shell Layers.
Control –the part of the system that initiates or stops the harmful function.	External pressure factor (predators, rocks etc.).

10.2.6 List of Potential Fields for Su-Field Analysis

According to the Su-Field model, every function can be described by an interaction of two substances in a given field. The field provides the source of energy to perform the function. List of potential fields for Su-Field analysis is presented in Table 10.9.

Table 10.9 List of potential fields for Su-field analysis

	Fields	Interaction including
M	Mechanical	Gravitation, collisions, friction, direct contact
		Vibration, resonance, shocks, waves
		Gas/fluid dynamics, wind, compression, vacuum
		Mechanical treatment and processing
		Deformation, mixing, additives, explosion
A	Acoustic	Sound, ultrasound, infrasound cavitation
T	Thermal	Heating, cooling, insulation, thermal expansion
		Phase/state change, endo-exo-thermic reactions
		Fire, burning, heat radiation, convection
C	Chemical	Reactions, reactants, elements, compounds
		Catalysts, inhibitors, indicators (pH)
		Dissolving, crystallization, polymerisation
		Odor, taste, change in color, pH, etc.
E	Electric	Electrostatic charges, conductors, insulators
		Electric field, electric current
		Superconductivity, electrolysis, piezo-electrics
		Ionization, electrical discharge, sparks
M	Magnetic	Magnetic field, forces and particles, induction
		Electromagnetic waves (X-ray, Microwaves, etc.)
		Optics, vision, color/translucence change, image
I	Intermolecular	Subatomic (nano) particles, capillary, pores
		Nuclear reactions, radiation, fusion, emission, laser
		Intermolecular interaction, surface effects, evaporation
C	Biological	Microbes, bacteria, living organisms
		Plants, fungi, cells, enzymes

Adapted with permission from the book "Improve your thinking: Substance-Field analysis", p. 17 [206]

10.2.7 *The Ideality Framework for Sustainability Analysis*

The ideality framework is based on TRIZ ideality law shown below.

$$\text{Ideality} = \frac{\text{All } \textbf{Useful} \text{ functions}}{\text{All } \textbf{Harmful} \text{ functions}} = \frac{\text{Benefits}}{\text{Costs}} \rightarrow \infty$$

General Remarks:

1. The purpose of this analysis is to identify sustainability design principles within the biological model (based on the correlation between ideality and sustainability).
2. System useful functions are benefits that the system provides.
3. System harmful functions are undesired costs of the system operation such as recourses, noise, waste, pollution etc.

10.2.8 The Ideality Patterns Table

The Ideality patterns table (Table 10.10) includes sustainability strategies and design principles that repeat in nature, extracted by the TRIZ ideality analysis of biological systems. The ideality patterns table together with the ideality framework

Table 10.10 Nature ideality patterns: strategies and design principles

	General strategy	Design principle	General example
Increase benefits	More functions	(1) Multifunctional design—enlarge the number of functions that are related to one structure by unification of system parts	Trees root's systems provide the function of channeling (nutrients and water) and the function of stability
	Stronger effect	(2) Intensify the interaction with the environment to achieve stronger effect of one function by: ✓ Repetition of elements ✓ Increased surface area	Repetition of pulmonary alveolus extends the gas exchange surface area, resulted in extended gas exchange
Reduce costs	Defensive strategy	(3) Reduction of disturbances such as friction, loads, turbulence by structures	Honeycomb structure reduces external loads effects
		(4) Decrease of surface area when it has harmful effect	Minimizing surface area of dessert leaves to reduce water loss by evaporation
	Opportunist strategy	(5) Usage of physical, chemical, geometrical and other effects and gradients as energy resources (saving energy costs)	✓ Using wind energy to disperse seeds ✓ Using temperature gradient to transfer heat
		(6) Adjustment of structure to function: structure provides the function (saving material costs)	The tubular structure of roots enables their penetration to soil
		(7) Transfer some functions to the supersystem (saving material costs). Using supersystem material resources	Puffer fish uses water from the supersystem to blow their body
	Prevent waste	(8) Synchronize system parameters to prevent waste	Synchronize seeds germination with environmental humid levels
		(9) Improve the conductivity of energy through the system to provide easier access and prevent waste of energy	The network structure of the tree's root system provides easier conduction and prevents loss of energy in the process
		(10) Give up redundant parts	Give up redundant filling material in bones provide lightweight bone and save material and energy

(Fig. 6.4) is considered as the ideality tool that supports the transfer of sustainability from biology to technology.

10.2.9 Transfer Platform: Analogy Comparison Components

These tools are suggested for the transfer stage and provide a platform to evaluate the similarities and differences between the biological solution and the biomimetic application. The comparison clarifies what is being transferred and what should be substituted in order to imitate the biological functional mechanism and sustainability aspects.

10.2.9.1 Compare the Complete Viable Biological and Biomimetic System Models

Analogy comparing components are presented in Fig. 10.3 and supported by Table 10.11. The designer should compare components 1–5 of the biological model and the biomimetic design concept model and identify differences and similarities.

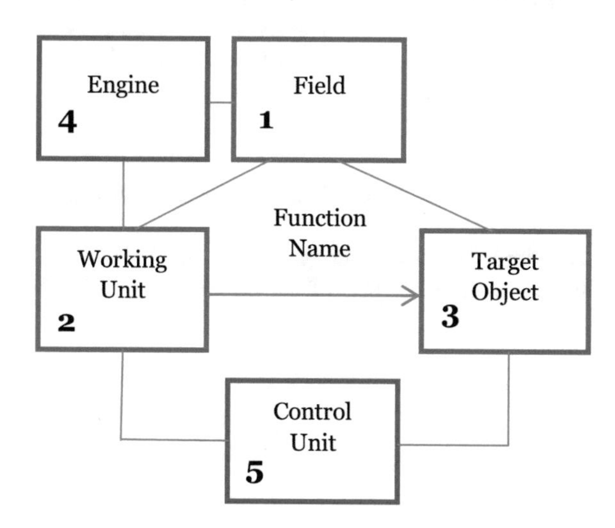

Fig. 10.3 Comparison between analogy components

Table 10.11 The complete viable analogy comparison table

		Biological model	Design concept	Remarks
1	Field			
2	Working Unit			
3	Target object			
4	Engine			
5	Control Unit			

Table 10.12 The ideality comparison table

General strategy	Design principle	Biomimetic concept	Biological model
Increase benefits			
More functions	(1) Multifunctional design		
Stronger effect of one function	(2) Intensify the interaction by: ✓ repetition of elements ✓ Increased surface area		
Reduce costs			
Defensive strategy prevent disturbances and harmful effects (saving the costs of disturbances)	(3) Reduction of disturbances		
	(4) Decrease of surface area		
Opportunist strategy Usage of available resources to save costs	(5) Usage of physical, chemical, geometrical and other effects and gradients as energy sources		
	(6) Adjustment of structure to function		
	(7) Transfer some functions to the supersystem		
Prevent waste for better usage of resources	(8) Synchronize system parameters to prevent waste		
	(9) Improve the conductivity of energy through the system		
	(10) Give up redundant parts		

10.2.9.2 Compare the Ideality (Sustainability) Strategies

The ideality comparison table (Table 10.12) is a template for comparing the ideality strategies and principles of the biological and biomimetic models.

Part IV
Experimentation

Following the introduction of the biomimetic design method by the method manual (Chap. 10), we introduce the experimentation process of the design method. This part corresponds to module 5 of the research model (Fig. 5.3). Experimentation serves to sharpen and refine the design method during the development process and during the validation of the final version of the design method. We experimented with the structural biomimetic design method in case studies and laboratory and field experiments. This part has accordingly two chapters:

Chapter 11: **Case Studies**: In this chapter, we present several case studies demonstrating the usage of the structural biomimetic design method.

Chapter 12: **Laboratory and Field Experiments**: In this chapter, we present two laboratory experiments performed by mechanical and industrial engineering students. One focuses on abstraction by the complete viable system model in relation to innovation. The other focuses on the ideality analysis in relation to sustainability. In addition, we present a field experiment performed at a medical company.

Chapter 11
Case Studies

In this chapter, we describe four case studies of biomimetic designs developed by two mechanical engineers, submitted as final projects of their M.Sc. degree in mechanical engineering. Their design process was documented and used as a mean to refine and modify the structural biomimetic design method, algorithm and tools.

First, the two mechanical engineers learned the design method from the authors. The authors introduced the knowledge bases of the structural biomimetic design method (Chaps. 7–9), and presented several case studies, previously developed by the authors, demonstrating the design process.

The two mechanical engineers were instructed to define a design challenge derived from their professional experience and solve it by the structural biomimetic design challenge (from a problem to biology). In addition, they were instructed to locate a biological system with a unique mechanism and develop a biomimetic solution following the structural biomimetic design method (from biology to an application). Altogether they developed six case studies. Four are presented here.

The two mechanical engineers reflected on their experience and elaborated on their difficulties and improvement suggestions during several periodic meetings with the authors. Thus for example, they reported on the extended time they devoted to search for biological systems. Though they did not measure that exact time, they support the evidence of the time consuming searching process. Their clarifications and modifications were valuable and used to refine the design algorithm and tools. All the process lasted 6 months and included six reflection meetings. The manual presented in Chap. 10 includes already their modifications. Examples for modifications performed due to their remarks:

- Number the ideality strategies and provide a short name to each one
- Simplify the complete viable model by reducing the transmission unit
- Improve definitions of the asymmetric structures
- Elaborate the third hierarchy of functions at the patterns table

Y.H. Cohen and Y. Reich, *Biomimetic Design Method for Innovation and Sustainability*, DOI 10.1007/978-3-319-33997-9_11

- Add comparison tables for the transfer stage
- Skip a stage of engine/brake identifications. Identify instead directly the structural pattern that determines if it is an engine or a brake.

In addition, they reflected on abandoned cases. Most of them were abandoned due to insufficient reference to explain the biological mechanisms.

11.1 From Biology to an Application

11.1.1 From Papilionaceae Seed to an Application

(a) Biological System

i. Encounter a biological system: The Papilionaceae seed [207]

The Papilionaceae seed resists water penetration and remains dormant in the soil for many years till the environmental conditions are adjusted for germination. The seed is kept dry by a humidity reduction mechanism that keeps the humidity levels inside the seed lower or equal to the humidity levels in the environment.

When the seed is detached from the plant, a scar remains in the detachment area. This scar is actually a hilum that functions as a unidirectional valve. Humidity can be released in one direction, from the seed outside, supporting the desiccation process of the seed. When the humidity levels in the environment are higher than the humidity levels within the seed, the hilum remains closed. When the humidity levels in the environment are lower from the humidity levels within the seed, the hilum opens and releases the humidity from the seed till it is equal to the humidity level in the environment.

The hilum itself is an asymmetric oval-shaped scar about 1 mm that is lying in a depression on the seed envelope. In the long axis of the hilum there are two epidermis layers. Each one is constructed of pillar cells vertically to the epidermis layer (Fig. 11.1). An impermeable tissue separates the external and internal epidermis layers. As a result, the external epidermis layer can react only to the humidity levels in the environment whereas the internal layer reacts to the humidity levels within the seed.

The pillar cells that construct the epidermis layers react to humidity by absorbing or desiccating humidity. When a pillar cell absorbs humidity it is swelled and when it releases humidity it is shrunk. The change in volume in these pillar cells creates a tension and loose movement as a result of the exerted hygroscopic pressure.

The pillar cells that define the height of the epidermis are distributed asymmetrically as the pillars are getting shorten when we approach the hilum (Fig. 11.2), having a concave shape. As a result, when the external epidermis cells absorb more humidity comparing to the cells in the internal epidermis layer (when the humidity level in the environment is higher than the humidity level within the seed), they swell and extract pressure that close the hilum. When the internal epidermis cells absorb more humidity comparing to the cells in the external epidermis layer (when

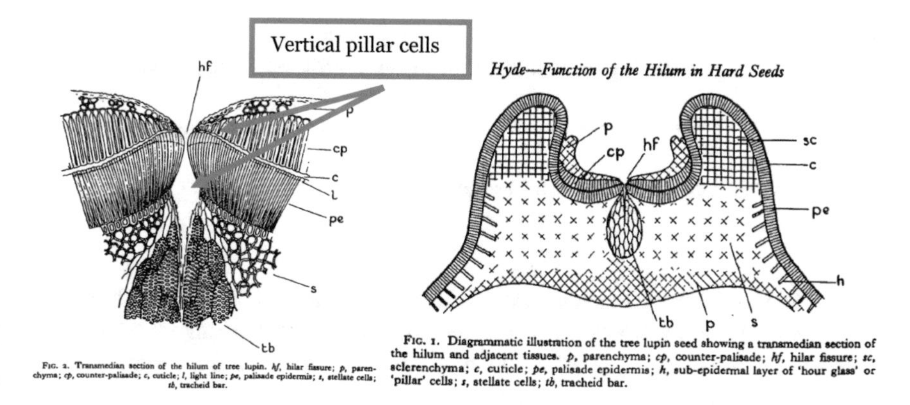

Fig. 11.1 The Papilionaceae seed: schematic figure of the scar and the hilum area. *On the right*, a schematic figure of the scar location on the seed envelope. *On the left*, a schematic description of the hilum area itself. Image adapted from [207], reproduced with permission of the copyright owner

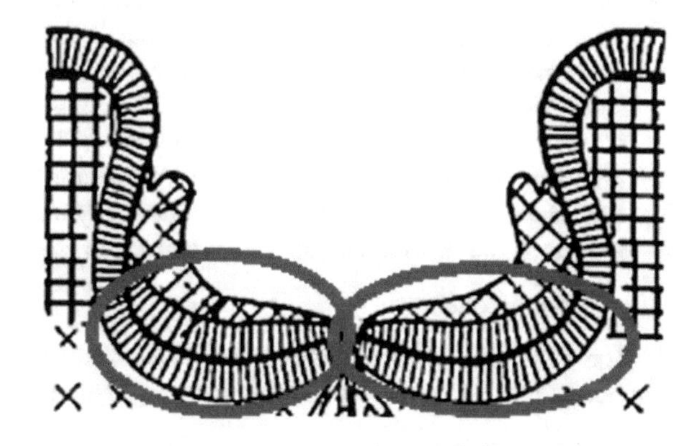

Fig. 11.2 The Papilionaceae seed hilum: pillar cells. The pillar cells that define the height of the epidermis are distributed asymmetrically as the pillars are getting shorten when we approach the hilum. Image adapted from [207], reproduced with permission of the copyright owner

the humidity level within the seed is higher than the humidity level outside the seed), they swell and extract pressure that opens the hilum.

(b) Bridging to technology

Identify possible analogical applications: Humidity devices (valves, controllers, and meters). Formulate the main function to be studied: Humidity levels regulation.

(c) Abstraction

i. Identify system parts

Supersystem	Papilionaceae plant, Humidity
System	Papilionaceae seed humidity regulation system

Subsystem Seed, seed envelope, seed scar (hilum)
Hilum: impermeable tissue, asymmetric epidermis layer (concave shape)
Epidermal layer: pillar cells swelled/shrunk (time asymmetry)

ii. Identify generic structures and related functions

Brakes

- *Container*—The seed envelope **contains** the seed as well as the inner humidity.
- *Layers*—The seed envelope is constructed of two epidermis layers. There is an impermeable tissue layer between them. It **protects** the seed from being exposed to humidity.

Engines

- *Repeated tubes*—Each epidermis layer is constructed of pillar cells that **regulate** the humidity levels within the seed.
- *Time Asymmetry*—Each pillar cell demonstrates time asymmetry as it shrinks or swells, **changing** its dimensions in time. This asymmetry is related to the movement (opening) of the hilum, as the swelling creates a hygroscopic pressure that **opens or closes** the hilum.
- *Geometric Asymmetry*—The pillar cells are distributed asymmetrically, as the pillars are getting shorter when we move towards the hilum. The height of each one of the epidermis layers is distributed asymmetrically. This asymmetry is related to the movement (**opening-closing**) of the hilum.

iii. System models

The following system models (Fig. 11.3) were developed for each one of the structures that participate in the humidity regulation mechanism.

iv. System Behavior

The identified structures are integrated together to perform a humidity regulation mechanism according to following stages:

- The impermeable layer of the seed envelope protects the seed from being exposed to external humidity levels. It is located between the two epidermis layers. Consequently, the pillar cells of the external epidermis layer react to external humidity levels. The pillar cells of the internal epidermis layer react to internal humidity levels.
- When humidity levels outside are higher, the pillar cells of the external epidermis layer swell more than the internal cells, they change their form, and extract hygroscopic pressure that closes the hilum, due to the geometric asymmetry of the epidermis layer (concave shape). When humidity levels inside are higher, the pillar cells of the internal epidermis layer swell more than the

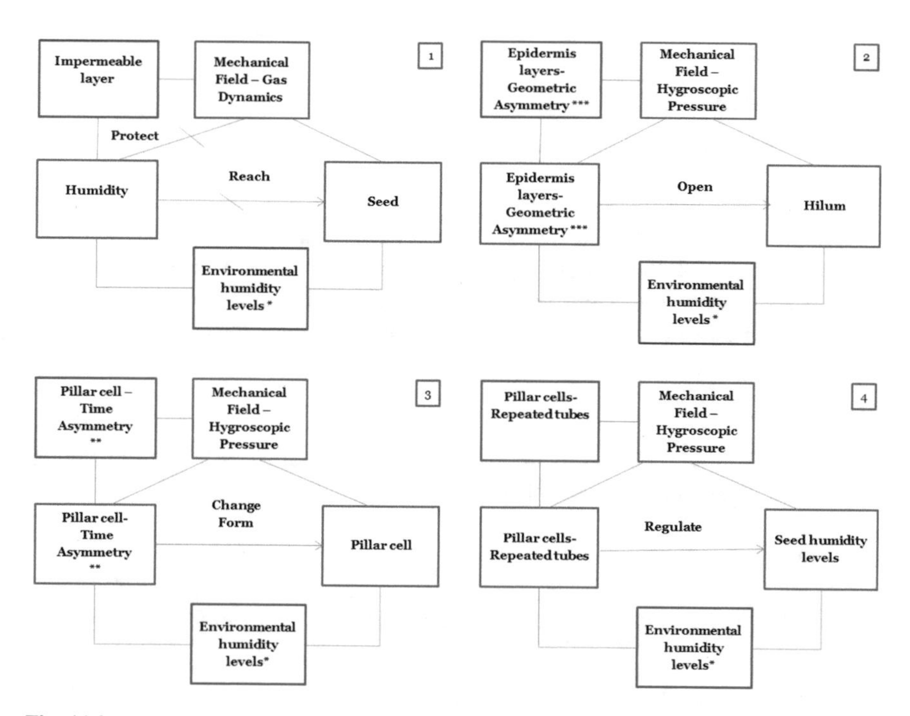

Fig. 11.3 The Papilionaceae seed humidity regulation system models. *When humidity outside the seed is lower than humidity inside the seed, the hilum opens and releases humidity. **Time Asymmetry: Pillar cell can swell and shrink. ***Geometric Asymmetry: Epidermis layers are shortened towards the hilum

external cells, change their form, and extract hygroscopic pressure that open the hilum, due to the geometric asymmetry of the epidermis layers (concave shape).
- In total, all pillar cells regulate the humidity level within the seed.

v. Ideality Analysis

Increase Benefits

Multifunctional designs:

- Seed envelope **contains** the seed and **protects** it against external humidity.
- Hilum connects the seed to the plant. When the seed fall it is a unidirectional valve.
- Hilum can protect against penetration of humidity from the environment when it is closed, and compare the humidity level to external humidity level when it is opened.

Stronger effect:

- Repetition of pillar cells intensifies the opening function.

Reduce Costs

Defensive strategy:

- Isolate the seed from humidity prevents decomposition.

Opportunistic Strategy:

- Using humidity gradient as energy source to open and close the hilum.
- The structure of the hilum (asymmetric epidermis layers) is adjusted to its opening function under certain conditions (unidirectional valve).
- Transferring functions to supersystem—using water particles (humidity) for the opening process.

Prevent Waste:

- Synchronization—synchronizing germination with humidity levels for optimal usage of seeds. Prevent loss of seeds (waste).

(d) Transfer

i. Identify an analogical design challenge for a chosen specific application

Unidirectional packaging valve: Packaging that provides low level of humidity to extend product shelf life. (The seed is analogical to the product; the seed envelope is analogical to the package). This packaging will include a unique unidirectional valve that prevents penetration of humidity. In addition, whenever the humidity levels outside are lower than the inner humidity levels, the valve will enable the release of humidity to the environment. Suppose the packaging of the product will be done in appropriate humidity conditions, the optimal humidity levels are maintained during the shipment stage. This packaging can be used for example in medical, food and electronic industries.

ii. Description of biomimetic design concept: Biomimetic packaging valve

The biomimetic valve has an asymmetric concave structure. It is made of two layers of matrix cells and impermeable layer between them (Fig. 11.4). In order to imitate the functionality of the pillar cells we use a matrix made of elastic material. Each cell of this matrix contains a material that reacts to increasing of humidity levels by swelling and to reduction of humidity levels by shirking. We design this material to extract pressure on the valve when the humidity level in the environment is higher or equal to the humidity level inside the packaging. In this case the valve is closed.

iii. Biomimetic System models

The biomimetic valve system models appear in Fig. 11.5.

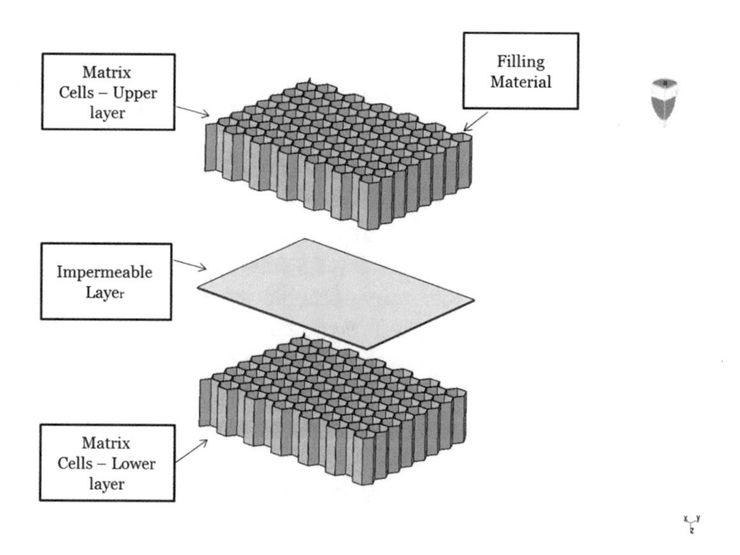

Fig. 11.4 Biomimetic unidirectional valve

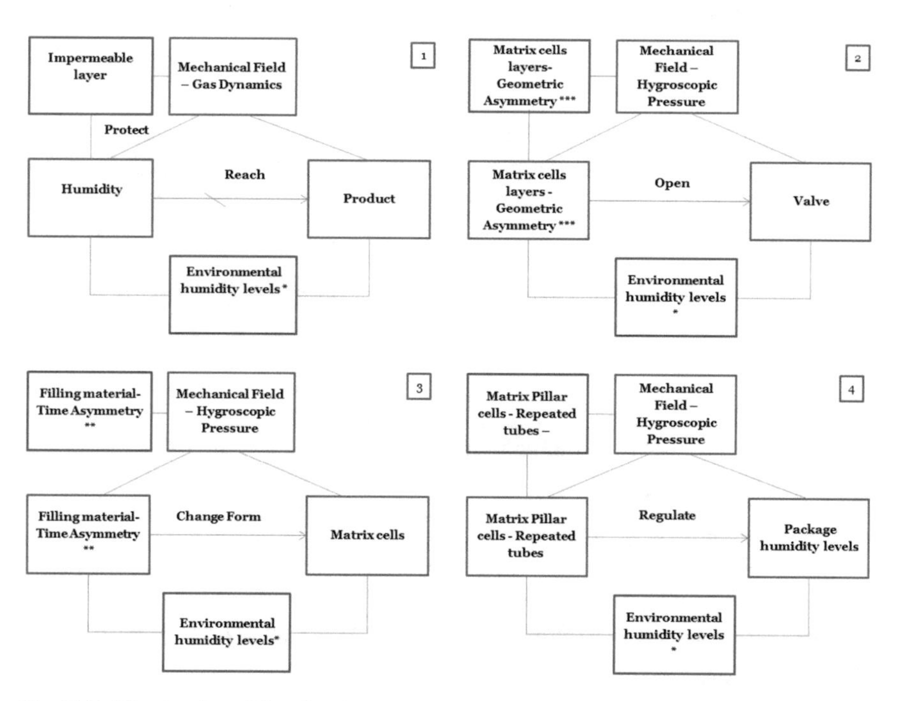

Fig. 11.5 Biomimetic unidirectional valve system models. *When humidity level outside the packaging is higher, the valve is closed and the product is protected by the impermeable layer. When humidity level outside the packaging is lower, the valve opens and releases humidity. **Time Asymmetry: Matrix filling material can swell and shrink. ***Geometric Asymmetry: Matrix layers are shortened towards the valve opening (concave shape)

iv. System Behavior

The identified structures are integrated together to perform the valve humidity regulation mechanism according to the following stages:

- The impermeable layer between the two matrix cell layers protects the product from being exposed to external humidity.
- When humidity level outside the package is higher, the filling material within the matrix pillar cells of the external layer swells more than the internal cells; they change their form and extract hygroscopic pressure that closes the valve, due to their concave shape (geometric asymmetry). When humidity level inside the package is higher, the filling material of the matrix pillar cells of the internal matrix layer swells more than the external cells; the cells change their form, and extract hygroscopic pressure that opens the valve, due to their concave shape (geometric asymmetry).
- In total, all matrix pillar cells regulate the humidity level within the package.

v. Ideality Analysis

Increase Benefits

Multifunctional designs:

- Package **contains** the product and **protects** it against the external humidity.
- Valve can protect against penetration of humidity from the environment when it is closed, and compare the humidity level to external humidity level when it is open.

Stronger effect:

- Repetition of matrix pillar cells intensifies the opening and closing function.

Reduce Costs

Defensive strategy:

- Isolate the product from humidity to prevent damage to product

Opportunistic Strategy:

- Using humidity gradient as energy source to open and close the valve.
- The concave structure of the valve (asymmetric matrix layers) is adjusted to its function of opening under certain conditions (unidirectional valve).
- Transferring functions to supersystem—using water particles to open valve.

Prevent Waste:

- Synchronize package with humidity level to extend life shelf and prevent waste.

vi. Compare biomimetic concept to biological solution (Tables 11.1 and 11.2)

Table 11.1 Papilionaceae seed case study: system models comparison

		Impermeable layer		Time asymmetric cells	
		Biomimetic concept	Biological model	Biomimetic concept	Biological model
1	Function	Prevent humidity	Prevent humidity	Change form	Change form
2	Field	Mechanical Field—Gas Dynamics	Mechanical Field—Gas Dynamics	Mechanical Field —Hygroscopic Pressure	Mechanical Field —Hygroscopic Pressure
3	Working unit	Humidity	Humidity	**Asymmetric filling material**	**Asymmetric Pillar cells**
4	Target object	**Product**	**Seed**	**Matrix cell**	**Pillar cell**
5	Engine/Brake	Impermeable layer	Impermeable layer	**Asymmetric filling material**	**Asymmetric Pillar cells**
6	Control unit	Environmental humidity levels	Environmental humidity levels	Environmental humidity levels	Environmental humidity levels
		Geometric asymmetric layers		**Repeated tubes/cells**	
		Biomimetic concept	**Biological model**	**Biomimetic concept**	**Biological model**
1	Function	Open Valve	Open Hilum	Regulate humidity	Regulate humidity
2	Field	Mechanical Field —Hygroscopic Pressure	Mechanical Field —Hygroscopic Pressure	Mechanical Field —Hygroscopic	Mechanical Field— Hygroscopic
3	Working unit	**Asymmetric concave matrix cells layers**	**Asymmetric concave Epidermis layers**	**Repeated tubes —Matrix Pillar cells**	**Repeated tubes—Pillar cells**
4	Target object	**Valve**	**Hilum**	**Package humidity level**	**Seed humidity level**
5	Engine	**Asymmetric concave matrix cells layers**	**Asymmetric concave epidermis layers**	**Repeated tubes —Matrix Pillar cells**	**Repeated tubes—Pillar cells**
6	Control unit	Environmental humidity levels	Environmental humidity levels	Environmental humidity levels	Environmental humidity levels

Differences are presented in bold

In Table 11.1, we compare the complete viable system models of the biological system and the biomimetic concept. We see that the biomimetic models keep the same functions, fields and control units as the biological models. The target object is different as expected, while the seed is replaced with the product, the hilum with the valve and the epidermal pillar cells with the matrix cells. The differences are mainly at the working units and the engines. The biological pillar cells are replaced with the artificial matrix cells and the filling material that can be swelled or shrunk in reaction to humidity. The substitutes keep the pillar cells geometric and time asymmetry.

Table 11.2 Papilionaceae seed case study: ideality strategies comparison

	Biomimetic concept	Biological model
Increase benefits		
More functions	Multifunctional Designs: 1. Package **contains** the product and **protects** it against external humidity 2. Valve can protect against penetration of humidity from the environment when it is closed, and compare the humidity level to external humidity level when it is open	Multifunctional design: 1. Seed envelop **contains** the seed and **protect** it against the external humidity 2. Hilum serves to connect the seed to the plant. When the seed fall it serves as a unidirectional valve 3. Hilum can protect against penetration of humidity from the environment when it is closed, and compare the humidity level to external humidity level when it is open
Stronger effect	Repetition of matrix pillar cells intensifies the opening/closing function	Repetition of epidermal pillar cells intensifies the opening/closing function
Reduce costs		
Defensive strategy	Isolate the product from humidity thus preventing decomposition	Isolate the seed from humidity thus preventing decomposition
Opportunistic strategy	Using humidity gradient as energy source to open and close the valve (hydrostatic engine)	Using humidity gradient as energy source to open and close the hilum (hydrostatic engine)
	The structure of the valve (asymmetric matrix layers) is adjusted to its function of opening under certain conditions (unidirectional valve)	The structure of the hilum (asymmetric epidermis layers) is adjusted to its function of opening under certain conditions (unidirectional valve)
	Transferring functions to supersystem—Using water particles (humidity) for the opening process	Transferring functions to supersystem—Using water particles (humidity) for the opening process
Prevent waste	Synchronizing package with humidity level to extend life shelf and prevent waste	Synchronizing germination with humidity levels for optimal usage of seeds. Prevent loss of seeds (waste)

In Table 11.2, we compare the ideality analysis of the biomimetic concept and the biological model. We see that Ideality strategies of the biomimetic concept preserved all relevant ideality strategies of the Papilionaceae seed humidity regulation system.

(e) Summary

The biomimetic design concept of a unidirectional valve is imitating the humidity regulation mechanism of the Papilionaceae seed almost completely. The design concept is based on the four core structures of the Papilionaceae seed and their behavior: Layers—array of two epidermis layers and impermeable layer between them; Time asymmetric cells that can shrink and swell; Geometric asymmetric concave valve; and finally the Repeated tubes of the whole array.

In order to imitate the humidity regulation mechanism there was a need to replace the following system components: Epidermis cells with the matrix cell but the filling material is the factor that shrinks and swells and not the matrix cell themselves, as happens in the biological system.

Ideality strategies of the biomimetic concept preserved all relevant ideality strategies of the Papilionaceae seed humidity regulation system. Multifunctionality of the hilum that served both a connecting tissue to the plant and a humidity valve is irrelevant to our discussion.

11.1.2 From Lizard Tail Autotomy to an Application

(a) Biological System

i. Encounter a biological system: Lizard Tail Autotomy [208]

Some lizards have the ability to voluntary shed their tail as a defensive mechanism against predators. The lizard can shed its tail when it is grasped by a predator or when it feels threatened. The detached tail flutters, attracts the predator's attention and let the lizard escape safely. A new tail is growing up within weeks to months.

The shedding of the tail always occurs in a distinct area of a fracture plane. This area is identified with "score lines" that create a weakness area. During the detachment of the tail, the fracture plane opens like a zipper, creating a "crown" of zipper protrusions (Fig. 11.6a, b).

This shedding mechanism needs to be rapid and therefore it is not based on chemical process of proteolysis, but on a mechanical structural process. The two

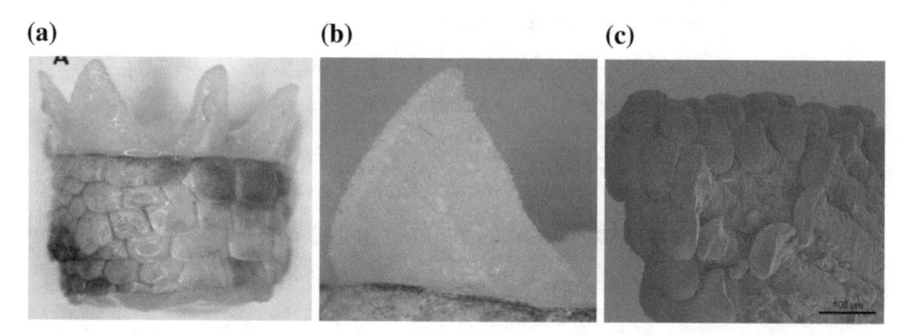

Fig. 11.6 End of lizard tail after the shedding. Adapted from [208] under CC Attribution License. **a** End of lizard tail after the shedding—the "zipper" crown structure. **b** Zoom on one zipper protrusion. **c** Scanning electron microscopy of one "zipper" protrusion: Muscles terminals have a "mushroom-shaped" structure

parts of the tail, before and after the fracture plane, are attached by biological adhering forces. The micro-structure in the fracture plane area enables the enlargement of the surfaces area when the tail is intact, and reduces the contact area rapidly during the shedding.

Before the shedding, microstructures at the terminal end of the muscle fibers have a shape of rod that enlarges the surface area and facilitates the biological adhesion. During the shedding, the lizard contracts the muscles around the fracture plane, and as a result, the shape of the rods is changed into "mushroom-shaped" structures (Fig. 11.6c). Consequently, the interaction area of the biological adhesion forces at the fracture plane is lowered, facilitating autotomy. The muscle contractions are also likely to facilitate removal of the skin and muscles to complete the release of the tail.

(b) Bridging to technology

Identify possible analogical applications: industrial or medical procedures that require a rapid controlled detachment of segment. For example: detachment of clutch, ejection seat, or a device that is entered by medical catheter and needs to be released at a given place.

Formulate the main function to be studied: rapid controlled detachment of a segment.

(c) Abstraction

i. Identify system parts

Supersystem	Predators, ground
System	Lizard tail autotomy system
Subsystem	Lizard brain
	Lizard Tail: skin, muscles, fracture plane, tail segments before and after the fracture plane
	Fracture plane: "score lines", zipper protrusion
	Zipper protrusions micro structure: terminal end of the muscle fibers —rods/"mushrooms" shape

ii. Identified generic structures and related functions

Brakes

- *Container*—The tail skin **contains** the tail organs such as muscles and blood vessels.

Engines

- *Fracture plane: Material asymmetry*—the tail contains "score lines" at distinct horizontal fracture plane. The "score lines" create an area that is weakened relatively to adjacent area without these "score lines". As a result, the strength of the tail is not equal along the tail. This asymmetry in the material strength defines the **cutting** area.
- *Repeated zipper protrusions*—the fracture area is opened like a zipper crown with several protrusions. These protrusions are related to the **attachment and detachment** of the tail segments, before and after the fracture plane.
- *Repeated rods/"mushrooms" (micro-structure)*—Each zipper protrusion is constructed of repeated terminal ends of muscle fibers. Before the shedding these terminal muscles have a rod shape. After the shedding, they have a "mushroom" shape. These protrusions are related to the biological adhesion of the two parts of the tail at the fracture area.
- *Time asymmetry*—the shape of each rod (terminal end of muscle) can be **changed** into a "mushroom" shape, demonstrating different shapes during time. The change of shape reduces the contact area and as a result reduces the adhesion forces, facilitating the tail shedding.

iii. System models

The following system models were developed for each one of the structures that participate in the Lizard tail autotomy system (Fig. 11.7). The container structure is general and is not modeled.

iv. System Behavior

The identified structures are integrated together to perform the lizard tail shedding mechanism according to following stages:

1. Regularly, when there is no threat, the lizard tail is intact. The repeated zipper protrusions in general and the microstructure of the repeated muscles rods in particular sustain the enlargement of the surface interaction area and the biological adhering forces. The two segments of the tail remain attached.
2. In case of a predator or threat: Microstructures at the terminal end of the muscle fibers contain many rods. Each rod changes into a mushroom shape, reducing the contact area of the muscles and reducing their adhesion. As a result, the adhesion forces are reduced and the tail segments are detached. This change process occurs either when a predator grasps the tail or when the lizard contracts the muscles.
3. The detachment occurs at the weakened of the fracture plane, where the material asymmetry defines the cutting area.

v. Ideality Analysis

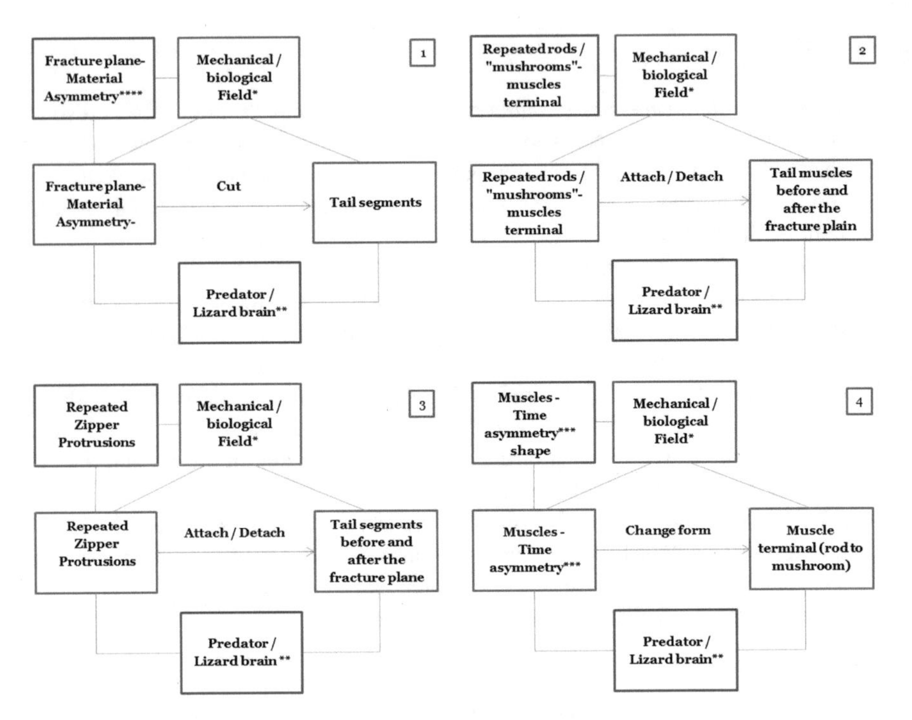

Fig. 11.7 Lizard tail autotomy system models. *Biological Field—when shedding without a predator force. Mechanical Field—in case of predator grasp. **A predator can apply pressure on the tail that initiates the detachment process. In case of threat without a physical touch of a predator, the lizard brain commands the muscles to contract and initiate the detachment process. ***Muscles can contract or release ****Fracture plane is identified with material asymmetry as it is weakened relative the rest of the tail

Increase Benefits

Multifunctional Design:

- The two segments of the tail can be both attached and detached.

Stronger effect:

- Quicker adhesion/cut as a result of the **repetition** of rods/mushrooms
- Quicker adhesion/cut as a result of a change of the contact area. Rod shape provides extended contact area and better adhesion. Mushroom shape provides reduced contact area and intensifies the cut.

Reduce Costs

Defensive strategy:

- Tail shedding to distract predators.

Opportunistic strategy:

- Using available energy sources: Using the predator's mechanical energy to change the rod shape and cut the tail (In case of a predator and not a general threat).
- Adjustment of structure to function: The zipper protrusions at the macro level and the repeated rods at the micro level are adjusted to the adhering and cutting functions.
- Adjustment of structure to function: The material asymmetry at the fracture area defines a weakened area and adjusted for a rapid cut.

Prevent Waste:

- Synchronizing the tail shedding with the presence of threat/predator.
- Give up redundant parts: the "score lines" within the fracture area give up redundant material.

(d) Transfer

i. Identify an analogical design challenge for a chosen specific application

Medical catheter for leading and releasing a medical device at a given place, such as a medical plug.

ii. Description of biomimetic design concept: Leading and releasing system for a medical sealing plug.

The plug is entered by a catheter to an area in order to seal a hole. Plug dimensions are adjusted to the geometric shape and dimensions of the hole. The plug is being fixed to the hole by a combination of pressure tolerance and biological glue. The plug is attached to the catheter at a fracture plane by normal (perpendicular) forces controlled by an electromagnet. Between the plug and the catheter there is a fracture plane identified by a zipper protrusions structures. The zipper protrusions are located at the end of the catheter and at the end of the plug, so they can be attached to each other in the zipper structure parallel surfaces. The catheter end and the plug have elastic properties. Every catheter protrusion's wall is embedded with a ferromagnetic substance that can be remotely activated for its magnetic properties, as in electromagnets. By arranging a different magnetic polarity for every protrusion, one can construct a binary array of positive and negative pole protrusions surface, as demonstrated in Fig. 11.8.

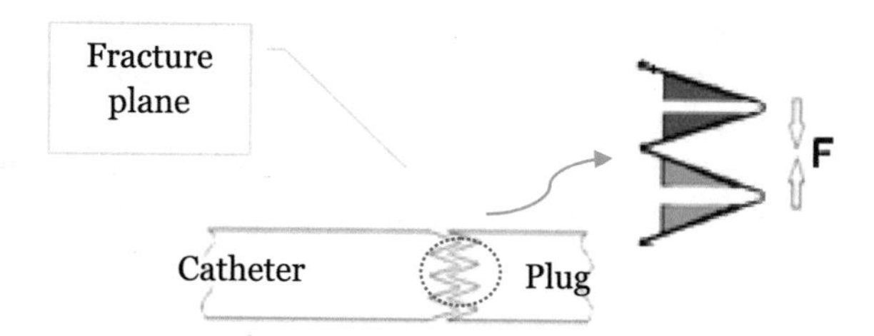

Fig. 11.8 Biomimetic leading and releasing system for sealing plug

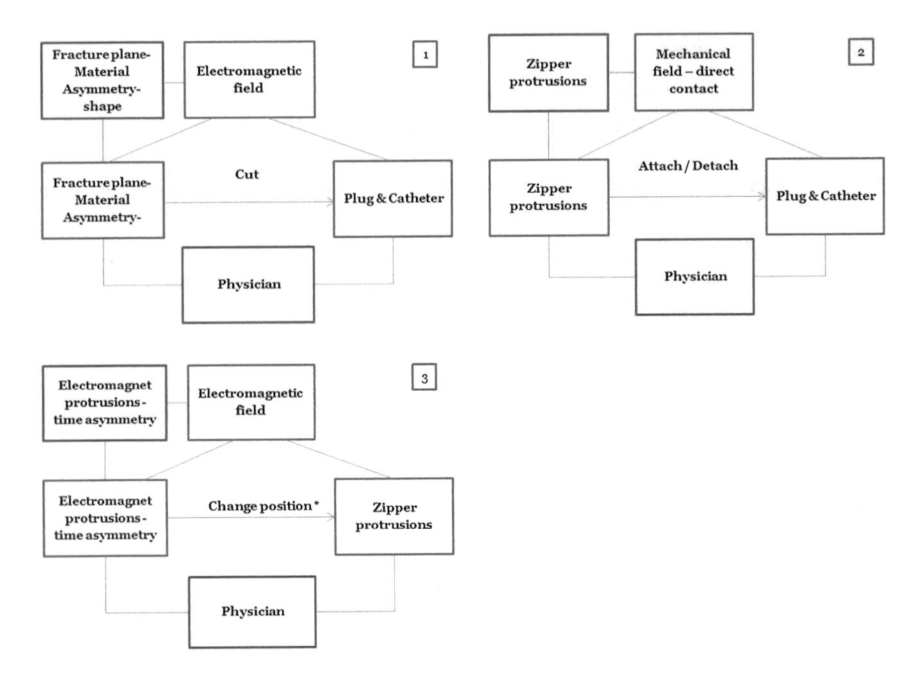

Fig. 11.9 Biomimetic sealing plug leading and releasing system models *Pulled or pushed

Activating the catheter's magnetic fields array will cause every protrusion on its surface to be swollen and every depression to be smaller. This will introduce normal (perpendicular) forces on the contact plane and will not allow the plug to be detached. Deactivating the magnetic fields will allow the plug to detach from the catheter.

iii. Biomimetic system models

We focus on the attachment and detachment of the plug as this is the analogical part to the biological system. System models of the biomimetic sealing plug appear in Fig. 11.9.

iv. System Behavior

The identified structures are integrated together to perform the plug releasing mechanism according to following stages:

- The ferromagnetic substance creates a magnetic polarity between the catheter's crowned protrusions.
- When the electromagnetic field is activated, the protrusions are swollen introducing normal (perpendicular) forces on the contact plane. The plug is attached to the catheter.
- Deactivating the magnetic fields abolish the normal forces and allow for the plug to detach from the catheter.

v. Ideality Analysis

Increase Benefits

Multifunctional Design:

- The system leads and releases the plug.

Stronger effect:

- Quicker adhesion/cut as a result of the zipper structure along the plug that increases the contact area between the plug and the catheter.
- Quicker adhesion/cut as a result of a change of the contact area due to change of the position of the zipper protrusions.

Reduce Costs

Defensive strategy:

- Decrease the danger to the patient and ease the operation as a need for an additional cutting device to cut the plug is prevented.

Opportunistic strategy:

- Adjustment of structure to function: The zipper protrusions are adjusted to the adhering and cutting functions.
- Adjustment of structure to function: The material asymmetry at the fracture area defines a weakened area and adjusted for a rapid cut.

Prevent Waste:

- Give up redundant parts: give up the need of a cutting device.
- Decreasing procedure costs due to decreasing of procedure time in relation to a state that requires an additional cutting device.

vi. Compare biomimetic concept to biological solution (Tables 11.3 and 11.4)

In Table 11.3, we compare the complete viable system models of the biological system and the biomimetic concept. The biomimetic models keep the same functions, while the tail segments are analogical to the plug and catheter segments of the leading system. We see that the biological field is replaced with an electromagnetic field as a source of energy for the cutting function. The electromagnet protrusions that change their shape in time (time asymmetry) imitate the muscles and function as the engine and the working unit of the system. The biological control of the lizard brain or predator is naturally replaced by the physician who manages the operation.

Table 11.4 compares the ideality analysis of the biomimetic concept and the biological model. We see that the biomimetic concept incorporates most of the ideality strategies demonstrated by the biological model, but we lose one degree of ideality by replacing the mechanical field (in case of a predator and not a general threat) with electromagnetic field.

Table 11.3 Tail Autotomy case study: system models comparison

		Fracture plane—Material asymmetry		Repeated zipper protrusions	
		Biomimetic concept	Biological model	Biomimetic concept	Biological model
1	Function	Cut	Cut	Attach/detach plug and catheter	Attach/detach tail segments
2	Field	**Electromagnetic and mechanical field**	**Mechanical/biological field**	**Mechanical field—Friction**	**Mechanical/biological field**
3	Working unit	Fracture plain—Material Asymmetry	Fracture plane—Material Asymmetry	Repeated zipper structure	Repeated zipper structure
4	Target object	**Plug and Catheter**	**Tail segments**	**Plug and Catheter**	**Tail segments**
5	Engine/Brake	Fracture plane—Material Asymmetry	Fracture plane—Material Asymmetry	Repeated zipper structure	Repeated zipper structure
6	Control unit	**Physician**	**Predator/Lizard brain**	**Physician**	**Predator/Lizard brain**

		Repeated rods/mushrooms		**Muscles Time asymmetry**	
		Biomimetic concept	**Biological model**	**Biomimetic concept**	**Biological model**
1	Function	Not transferred	Attach/Detach tail segments	Change position	Change form
2	Field		Mechanical/biological field	**Electromagnetic field**	**Mechanical/biological field**
3	Working unit		Repeated rods/mushrooms	**Electromagnet—Protrusions time asymmetry**	**Muscles time asymmetry**
4	Target object		Tail segments	**Zipper protrusions**	**Muscle terminal (rod to mushroom)**
5	Engine/Brake		Repeated rods/mushrooms	**Electromagnet—Protrusions time asymmetry**	**Muscles Time asymmetry**
6	Control unit		Predator/Lizard brain	**Physician**	**Predator/Lizard brain**

Differences are presented in bold

Table 11.4 Tail autotomy case study: ideality strategies comparison

	Biomimetic concept	Biological model
Increase benefits		
More functions	Multifunctional design: The plug can be both attached and detached	Multifunctional design: The two segments of the tail can be both attached and detached
Stronger effect	Quicker adhesion/cut as a result of the zipper structure along the plug The repeated protrusions of the zipper enlarge the contact area between the plug and the catheter	Quicker adhesion/cut as a result of the repetition of rods/mushrooms Quicker adhesion/cut as a result of a change of the contact area Rod shape provides extended contact area and better adhesion Mushroom shape provides reduced contact area and intensifies the cut
Reduce costs		
Defensive strategy	Decreasing the danger to the patient as a need for an extra cutting device to cut the plug is prevented	Tail shedding distracts predators
Opportunistic strategy		Using available energy sources: Using the predator's mechanical energy to change the rod shape and cut the tail (In case of a predator and not a general threat)
	Adjustment of structure to function: The zipper protrusions are adjusted to the adhering and cutting functions	Adjustment of structure to function: The zipper protrusions at the macro level and the repeated rods at the micro level are adjusted to the adhering and cutting functions
	Adjustment of structure to function: The material asymmetry at the fracture area defines a weakened area adjusted for a rapid cut	Adjustment of structure to function: The material asymmetry at the fracture area defines a weakened area and adjusted for a rapid cut
Prevent waste	Synchronizing the plug dimensions with hole dimensions	Synchronizing the tail shedding with the presence of thereat/predator
	Give up redundant parts: give up the need of a cutting device	Give up redundant parts: the "scores" within the fracture area demonstrate give up on redundant material
	Decreasing procedure costs due to decreasing of procedure time in relation to a state that requires an additional cutting device	

(e) Summary

The biomimetic concept uses the biological shedding principles partially. The biomimetic concept imitates the zipper structure and the fracture plane, while the two micro-structures of the repeated rods/mushrooms and muscles (time asymmetry) are replaced by an electromagnet that provides the ability of changing the position of the zipper protrusions (pulls to each other or pushed from each other) and consequently, changing their normal force that attach them together, analogical to the biological adhesion. Here we suggest using an electromagnet but other mechanism based on pressure may also work.

The biological field that contracts the muscles is replaced with an electromagnetic field that pushes and pulls the catheter zipper protrusions. The operator of the medical procedure (physician) functions as the control unit and replaces the lizard brain/predator.

The biomimetic concept incorporates most of the ideality strategies demonstrated by the biological model. By replacing the mechanical field (in case of a predator and not a general threat) with electromagnetic field we lose one degree of ideality of using available energy sources. However, we offer another ideality strategy that is not relevant to the biological case—reduction of procedure time and costs and providing better protection to the patient compared to a state of using a separate cutting tool.

11.2 From a Problem to Biology

11.2.1 Dynamic Screen Protector

(a) Problem Definition

i. Identify design challenge: Dynamic screen protector.

Challenge description: Most of the electronic devices use screens for watching or operation (touch screens). One of the major factors that damage these electronic devices in general and the screens in particular is mechanical damage resulting from the device falling down. The common protection against fall today is a screen cover.

Screen covers have two major categories: The first is a wallet cover. This cover wraps the whole device. The user must open the cover before the use and close it back after the use. The relative size of the device is large due to the permanent cover. The second is a "partial" cover that wraps permanently only the back and sides of the device. This type does not provide full protection when the screen gets a direct hit, however the screen is accessible for watching or operation all the time.

ii. Identify main function of the design challenge

Protect the screen against mechanical damage without affecting the accessibility of the screen. There is a need that the screen will be protected during fall and accessible during use.

iii. Analyze the design space and identify possible design paths

The designers (M.Sc. students) observed several alternatives of possible design paths. The possibilities are presented by the following function-means tree (Fig. 11.10). The chosen design path is presented with dashed arrows.

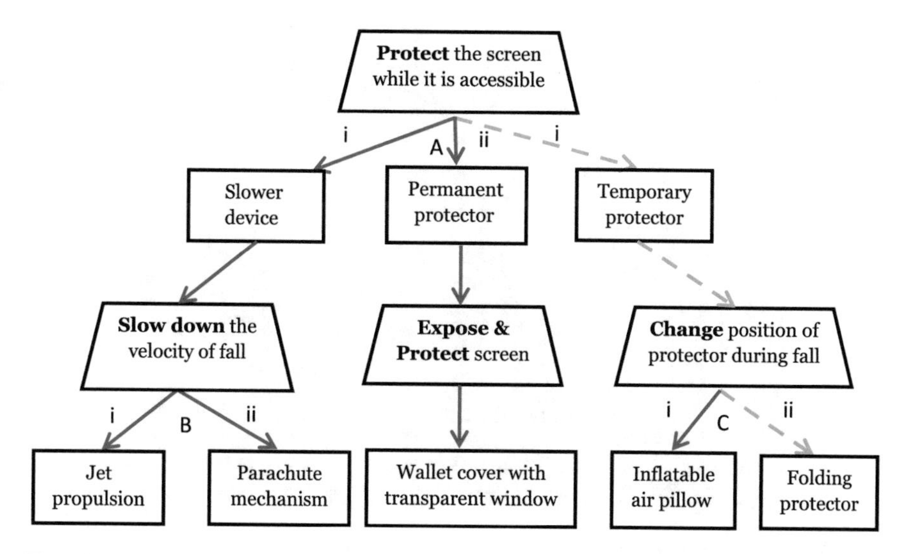

Fig. 11.10 Screen protector design challenge: function-means tree. Explanation of the function-means tree branching: **A** Separation by means and time—(*i*) mean during fall (*ii*) mean during regular use. **B** Separation by means and space—(*i*) mean located below the product—Jet propulsion (*ii*) mean located above the product—Parachute. **C** Separation by means—(*i*) mean that changes the volume of the cover (*ii*) mean that changes the position of the cover

The design path of slower device (Ai) is disqualified as the application is not simple. In the case of the jet propulsion (Bi) it also requires significant amount of energy.

The design path of permanent protector with transparent window (Aii) is disqualified as it disrupts the usage experience to watch or operate, in case of a touch screen. The design path of dynamic screen protector with inflatable air pillow (Ci) is disqualified as it requires a significant amount of energy to blow rapidly the pillow during the device fall. Eventually, we chose the design path of folding protector (Cii).

(b) Bridging to technology

Formulate the design challenge by generic functions and identify related structures

- The solution will **protect** the screen against mechanical damage—layers, intersected layers, cylinder, helix
- The solution will **change the position of protector** from full cover to full exposure—Asymmetric structure

(c) Biological System

We searched for a biological system that is identified with the above mentioned structure-function patterns (section b). We used the searching word of "**protect**"

Fig. 11.11 Pangolin scales. Image by Sandip Kumar, CC 3.0

and located the pangolin (Fig. 11.11) as a suitable role model in the "AskNature" [87] database, as it can **change the position** of its scales.

The pangolin has a flexible armor of horny keratin scales that overlap like shingles of a roof [87]. The scales cover the tail and the body. When the animal feels danger it moves its head into the stomach and rolls itself into a ball shape. Muscles are responsible for the rolling activity [209]. The scales may move one on another with partial overlap between the scales that enable the formation of a ball shape. The scales are sharp and can cut a predator if he tries to attack, providing extra defense. It is considered almost impossible to force the pangolin to unroll.

(d) Abstraction

i. Identify system part

Supersystem	Predators, soil, stones
System	Pangolin rolling into a ball system
Subsystem	Armor, muscles, Nerve system (brain)
	Armor: scales

ii. Identified generic structures and related functions

Brakes

- *Container*—The armor **contains** the pangolin organs.
- *Layers*—The overlapping scales **protect** the pangolin against mechanical loads.

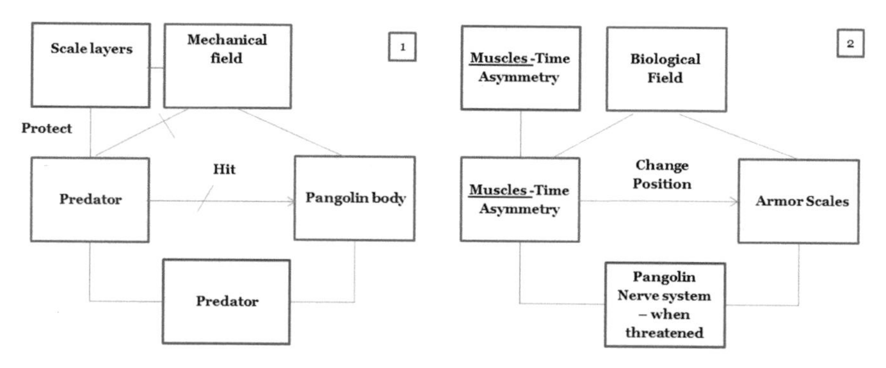

Fig. 11.12 Pangolin rolling into a ball system models

Engines

- *Muscles*—**Time Asymmetry**—Muscles can contract and release moving forward and backwards the scales, thus changing their position from a state of no overlap to a state of overlap.

i. System models

The system parts that are related to the pangolin rolling into a ball shape are shown in Fig. 11.12.

ii. System Behavior

- Regularly, when there is no threat (predator), the **armor scales** are flattened over the body, providing a reasonable defense against mechanical loads.
- In a state of threat (potential **predator**), the pangolin **nerve system** commands the **muscles** to contract. The muscles convert **biological energy** to mechanical motion and turn the pangolin body into a ball shape. The scales **change their position**, move and overlap accordingly. The pangolin is "locked" inside the armored ball.
- When the threat is gone, the pangolin nerve system commands the muscles to release. The muscles convert biological energy to mechanical motion and unroll the body and accordingly the scales. The pangolin gets back to his regular condition.

iii. Ideality Analysis

Increase Benefits

Multifunctional Design:

- The armor **contains** the inner organs and protects them.
- The armor may be flat or balled.

- The scales **protect** against mechanical loads by absorbing the blow and can hit back the predator by their sharpened edges.

Stronger effect:

- Repetition of elements—the repetition of scales enables turning into a ball shape. The more scales, the more "balled" is the shape and the protection to the body is extended by thicker armor in the overlapped area and stronger defense.

Reduce Costs

Defensive strategy:

- Reduction of external disturbances—the predator's threat is reduced by the balling mechanism.
- Reduce surface area—the surface area that is exposed to predators in a ball shaped is reduced relatively to unrolled shape.

Opportunistic strategy:

- Adjustment of structure to function—the scales structure enables their overlapping movement, turning into a ball shape and unrolled again.

Preventing Waste:

- Synchronizing system parameters to prevent waste—turning into a ball shape is synchronized with the presence of predators or threat, investing the energy in balling just when needed.

(e) Transfer

i. Describe biomimetic design concept: Dynamic screen cover for smart phones

The biomimetic concept offers protection to the screen **during fall**. An accelerometer, that is an integral part of each smart phone, identifies the fall. The shield is combined of two identical scale mechanisms located at the end of the smart phone as presented in Fig. 11.13. During the fall, these scales are opened and spread over the screen to protect it. In all other times, they are folded in the sides.

As there is a need for rapid reply, we will use a spring that can store and release elastic energy. The release of the spring is done by an electro mechanical component right after the fall recognition by the accelerometer. At the end of the fall the user can close back the scales by his hands, pushing back the spring.

ii. Biomimetic System Model—Are presented in Fig. 11.14
iii. Biomimetic System behavior

- During a regular use, the **cover scales** are closed at the sides of the device, leaving the screen accessible.
- In a state of fall, the **accelerometer** identifies the fall by the change of speed and activates the **electro mechanic component** that frees the **spring**.

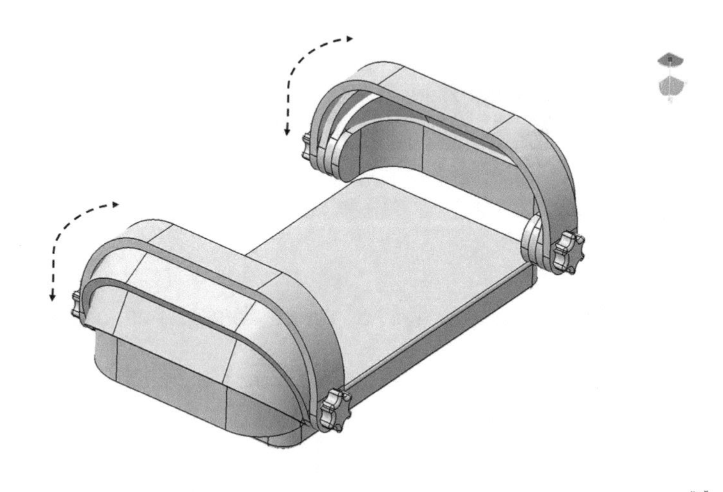

Fig. 11.13 Biomimetic dynamic cover in an opened state. The *dashed arrows* indicate the direction of the scales opening and closing

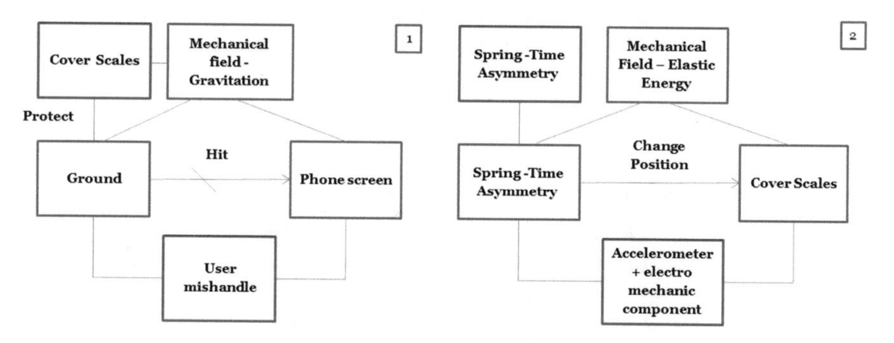

Fig. 11.14 Biomimetic dynamic cover system models

- The spring releases its elastic energy that is used to open the screen scales and cover the screen.
- After the fall, the user can push back the spring and the scales are located again at the sided of the device, leaving the screen accessible.

iv. Ideality Analysis

Increase Benefits

Multifunctional Design:

- The screen scales can be flat or balled.
- The dynamic scales protect against hit and access the screen for operation.

Stronger effect:

- Repetition of elements—the repetition of scales enables turning into a ball shape. The more scales, the more "balled" is the shape, the thicker protecting layer and better defense.

Reduce Costs

Defensive strategy:

- Reduction of disturbances: the risk of screen damages and repair costs is reduced.
- Reduce surface area—the surface area that is exposed to ground hit is minimized when the scales cover is opened.

Opportunistic strategy:

- Adjustment of structure to function—the scales structure enables to use the cover during fall and close it regularly.

Preventing Waste:

- Synchronizing system parameters to prevent waste—the cover scales are synchronized with fall, leaving the phone accessible without disturbances regularly.

v. Compare biomimetic concept to biological solution (Tables 11.5 and 11.6)

Table 11.5 The screen protector case study: system models comparison

		Scale layers		Muscles-time asymmetry	
		Biomimetic concept	Biological model	Biomimetic concept	Biological model
1	Function	Protect	Protect	Change position	Change position
2	Field	Mechanical Field – Gravitation	Mechanical Field	**Mechanical Field – Elastic Energy**	**Biological Field**
3	Working unit	**Ground**	**Predator**	**Spring-Time Asymmetry**	**Muscles-Time Asymmetry**
4	Target object	**Phone screen**	**Pangolin Body**	Cover scales	Armor scales
5	Engine/Brake	Cover scales	Scales layers	**Spring-Time asymmetry**	**Muscles-Time asymmetry**
6	Control unit	**User mishandle**	**Predator**	**Accelerometer + electro mechanic component**	**Pangolin nerve system – When threatened**

Differences are presented in bold

Table 11.6 The screen protector case study: ideality strategies comparison

	Biomimetic concept	Biological model
Increase benefits		
More functions	Multifunctional design— ✓ The screen scales can be flat or balled. ✓ The dynamic scales protect against hit and access the screen for watching/operation	Multifunctional design— ✓ The armor **contains** the inner organs and protects them. ✓ The armor may be flat or balled. ✓ The scales **protect** against mechanical loads by absorbing the blow and can hit back the predator by their sharpened edges
Stronger effect	Repetition of elements—The repetition of scales enables turning into a ball shape. The more scales the more "balled" is the shape and the more distanced is the phone from ground—better defense	Repetition of elements—The repetition of scales enables turning into a ball shape. The more scales the more "balled" is the shape and the protection to the body is extended: thicker armor in the overlapped area and stronger defense
Reduce costs		
Defensive strategy	Reduction of external disturbances—The risk of screen damages is reduced and so are the repair costs	Reduction of external disturbances—Predator's threat is reduced by the balling mechanism
	Reduce surface area—The surface area that is exposed to ground hit is minimized when the scales cover are opened	Reduce surface area—The surface area that is exposed to predators in a ball shaped is reduced relatively to unrolled shape
Opportunistic strategy	Adjustment of structure to function—The scales structure enables to use them during fall and close them regularly	Adjustment of structure to function—The scales structure enables their overlapping movement, turning into a ball shape and unrolled again
Prevent waste	Synchronizing system parameters to prevent waste—The cover scaled are synchronized with fall, leaving the phone accessible without disturbances regularly	Synchronizing system parameters to prevent waste—Turning into a ball shape is synchronized with the presence of predators or threat, investing the precious energy in balling just when needed

In Table 11.5 we compare the biological and biomimetic complete viable system models. We see that the functions stay the same while the pangolin that needs to protect against a predator is analogical to the phone screen that needs protection against fall. In order to imitate to movement of the scales we need to replace the biological field with mechanical field of elastic energy while the spring replaces the muscles and function as the working unit. The control should be also changed and the accelerometer replaces the pangolin nerve system.

In Table 11.6 we compare the ideality strategies of the biomimetic concept and the biological model. The biomimetic concept uses most of the relevant ideality strategies. We didn't lose some degree of ideality during the transfer stage.

(f) Summary

The biomimetic concept is based on the balling process of the pangolin armor and imitated the scales structure and their ability to move and overlap on each other. Generally, the armor scales are analogical to the cover scales, and the predator is analogical to the possible user mishandle that lets to the phone falling down. In both cases there is a need to protect the pangolin/screen just in time, leaving the body unrolled or the screen accessible, when there is no risk.

In order to imitate the folding process there was a need to replaces the following system components:

- The engine and the working unit—the muscles are replaced with the spring however both of them have the ability to store and release energy. While the muscles use biological field, the spring uses mechanical field of elastic energy. The spring can react in milliseconds and translate the elastic energy into a mechanical motion rapidly.
- The function of the spring may be also achieved by the click beetle jumping mechanism (Sect. 8.1.1.3), that is based on a hinge between contracted muscles. Though the M.Sc. students didn't identify the click beetle in this relation, it is a good example that demonstrates how the function-means tree can lead the search of various analogies in nature and how pre-observed organism, such as the click beetle, becomes a part of a design tools box.
- The control unit—The pangolin nerve system that initiates the rolling process is replaced by the accelerometer that identifies the fall and activates the electro mechanic component that releases the spring. At the end of the fall the user can close back the scales by his hands, pushing back the spring.

The biomimetic concept uses most of the relevant ideality strategies. As the dynamic screen phone cover does not contain the phone but exists in its sides, the multifunctional design of the armor as a container and protector is irrelevant in the biomimetic case. Generally we may say that we didn't lose some degree of ideality during the transfer stage.

11.2.2 Parking Space Reducer

(a) Problem Definition

i. Identify design challenge: Parking space reducer

Challenge description: Many big cities suffer from a serious shortage of parking space. The root of the problem is a surplus of vehicles in a limited fixed space. Non-optimal usage of the current available parking space leads to a waste of precious parking space and worsen the problem. For example, in the city of Tel-Aviv there are 120 thousands of parking spaces. One parking is estimated to cost 100,000 Shekels [210]. If we find a better way to use even 1 % more of the

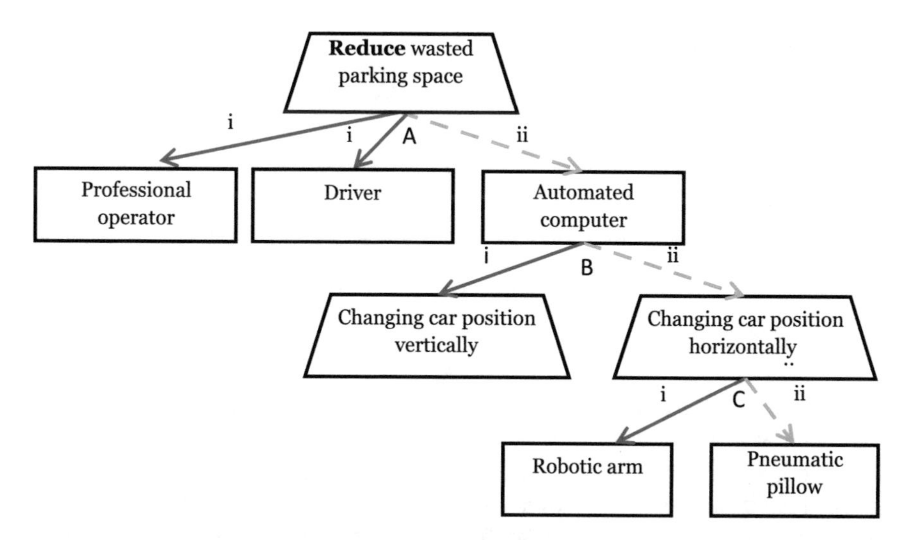

Fig. 11.15 Parking space reducer design challenge: function-means tree. Explanation of the tree branching: **A** Separation by means and space—(*i*) External mean—by a human operator (*ii*) Internal mean—by an automated computer. **B** Separation by functions and space—(*i*) Change car position vertically (setting up on front) (*ii*) Change car position horizontally (forwards and backwards). **C** Separation by means—various means to move the car forward and backwards

current parking space, 120 M Shekel could be saved, 10 % better usage will save 1.2 B and so on. Today, there is no major system that enables an optimal usage of parking spaces in big cities.

ii. Identify main function of design challenge

Parking space optimization by reduction of wasted parking space.

iii. Analyze the design space and identify possible design paths

The design space is defined as vehicles that park by sidewalks. We observed several alternatives of design path. They are presented by the function-means tree in Fig. 11.15.

The chosen design path is presented with dashed arrows. The design path of human operator (Ai) is disqualified as it depends on human capabilities and attention. The design path of changing car position vertically (Bi) by setting it up is disqualified due to safety and quality problems. Setting up the car on front can cause leakage of liquids such as oil and petrol, and damage the car. The design path of changing car position forwards and backwards by a robotic arm (Ci) was disqualified due to possible damage of the car by the arm. Eventually, we chose the design path of changing car position forwards and backwards by pneumatic pillow system (Cii).

(b) Bridging to technology

Formulate the design challenge by generic functions and identify related structures

- The solution will **change the position** of vehicles in a given space—asymmetric structure.
- The solution will **contain and protect** vehicles in a given space—a container structure

(c) Biological System

We searched for a biological role model that is identified with the above mentioned structure-function patterns (section b). We consulted a zoologist who offered the idea of the snake swallowing mechanism, as it both contains the prey and move it forward. Some snakes can lead the complete prey in their body after they swallow it. The prey is moved by muscles that contract and release before and after the prey [211]. After the prey is moved from the jaws to the throat we identify undulation movements that push it towards the stomach. The bends of the movement exert bilaterally inward forces against the soft prey (Fig. 11.16). It is assumed that the same muscles operate both swallowing and locomotion as they activated in a similar timing.

(d) Abstraction

i. Identify system parts

Supersystem	Prey, soil, land cover (rocks, vegetation), competitors
System	The snake prey swallowing mechanism
Subsystem	Jaws, throat, digestion system, esophagus, nerves system, muscles

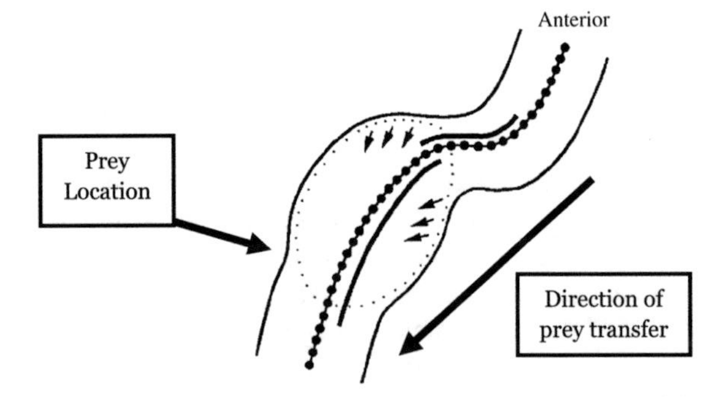

Fig. 11.16 The trunk of gopher snake. Muscle activity is indicated by the *filled bars* and the areas of presumed force exertion are indicated by the *arrows*. Image reproduced with permission from The Journal of Experimental Biology [211]

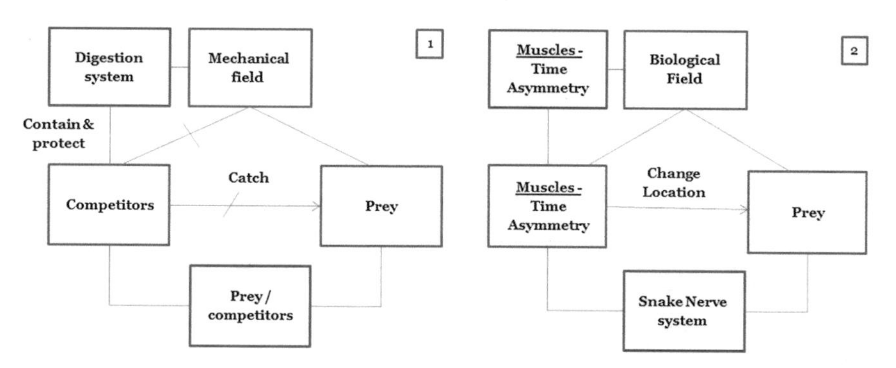

Fig. 11.17 The snake prey swallowing mechanism system models

ii. Identified generic structures and related functions

Brakes

- *Container*—The snake digestion system **contains** the prey

Engines

- *Muscles*—**Time Asymmetry**—Muscles can contract and release moving the prey in the digestion system.

iii. System models

We analyze the system parts in relation to the snake swallowing mechanism (Fig. 11.17).

iv. System Behavior

- When there is no prey, the digestive system is empty. The muscles are not activated.
- In a state of prey, the snake **digestive system contains the prey**. The snake **nerves system** commands the **muscles** to contract. The muscles convert **biological energy** to mechanical motion and move forward the prey in the digestive system.

v. Ideality Analysis

Increase Benefits

Multifunctional Design:

- The muscles that lead the prey are also used for the snake locomotion.
- The prey is contained in the body and kept away of competitors.

Reduce Costs

Defensive strategy:

- Reduction of external disturbances: the snake hide the prey against potential competitors

Opportunistic strategy:

- Adjustment of structure to function—the contraction of muscles location is adjusted to the required movement.
- Usage of available energy sources—the snake energy movement is used also for moving the prey.

Prevent Waste:

- Synchronization—swallowing movement is synchronized with locomotion movement—saving energy.

(e) Transfer

i. Describe biomimetic design concept—Parking space reducer

The system is constructed of a sleeve. The sleeve sides are made of air pillows that can be blow or emptied by a pneumatic system. By blowing and empting the air pillows before and after the vehicle, the vehicle can be moved, analogically to the prey movement by the muscles. However, opposing to the snake system where the prey movement is unidirectional, the vehicle can be moved forward or backward per need, according to the blow location. The blow creates a pressure that pushes the vehicle. If the blow is located ahead of the vehicle, it will move backwards and vice versa. Parking of vehicles is done always at the beginning of the sleeve. Vehicles are entered to the sleeve in a line and are ejected at the sides of the sleeve. The biomimetic parking reducer is presented in Fig. 11.18. Its system models are presented in Fig. 11.19.

ii. System models

iii. System Behavior

- When there is no vehicle, the parking system is empty. The pneumatic air pillows are not activated.
- When there is a vehicle, the computer system operator activates the air pillows to be blown and move forward the vehicle in the sleeve parking system. The vehicle is contained in the sleeve and is protected against vandalism, thieves, and climate hazards.
- When another vehicle comes, the air pillow is activated to push it exactly by the end of the previous vehicle. No space is left between the cars.
- If a vehicle is required to be ejected from the middle of the line, the air pillows are moved up and the vehicle is ejected to the side. The remaining apace is closed by the system pushing forwards or backwards the other vehicles in the line.

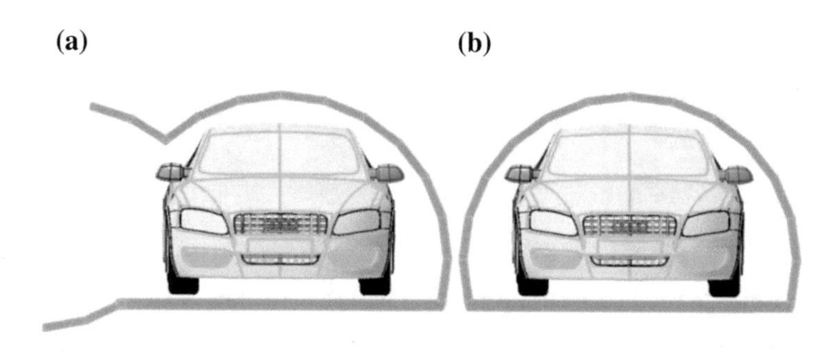

Fig. 11.18 Biomimetic Parking reducer: **a** Sleeve sides are opened to eject a vehicle. **b** Vehicle located inside the sleeve

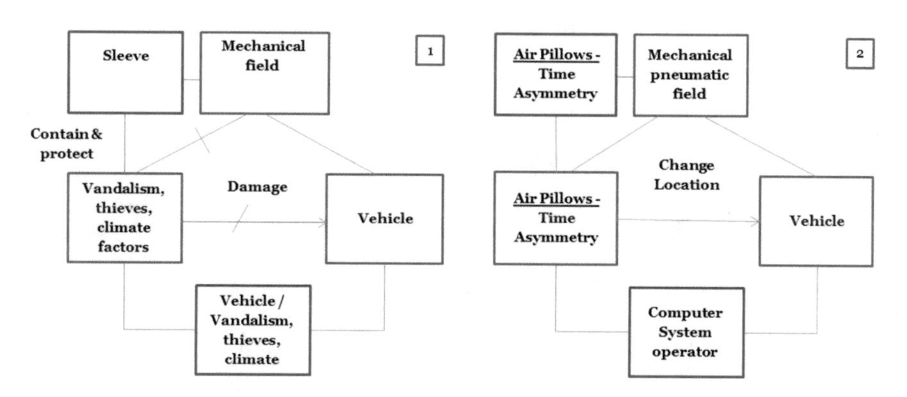

Fig. 11.19 Biomimetic parking reducer system models

iv. Ideality analysis

Increase Benefits

Multifunctional Design:

- The system contains and moves the vehicle.
- The sleeve protects the vehicle against thieves, vandalism and climate.

Reduction of Costs

Defensive strategy:

- The system prevents accidents during parking and saves repair costs.

Opportunistic strategy:

- Adjustment of structure to function—blowing of the air pillow is adjusted to the required movement.

Prevent Waste:

- Synchronization—the system can provide real time data on free parking space. Users can use these data and save time, costs and air contamination during the parking search.
- Give up redundant parts—the waste of parking space is prevented. As a result, the parking cost is reduced.

Table 11.7 Parking reducer case study: system models comparison

		Container		Muscles time asymmetry	
		Biomimetic concept	Biological model	Biomimetic concept	Biological model
1	Function	Contain and protect	Contain and protect	Change location	Change location
2	Field	Mechanical field	Mechanical field	**Mechanical pneumatic field**	**Biological field**
3	Working unit	**Vandalism, thieves, climate factors**	**Competitors**	**Air Pillows-Time asymmetry**	**Muscles-Time asymmetry**
4	Target object	**Vehicle**	**Prey**	**Vehicle**	**Prey**
5	Engine/Brake	**Sleeve**	**Digestion system**	**Air Pillows-Time asymmetry**	**Muscles-Time asymmetry**
6	Control unit	**Vehicle**	**Prey**	**Computer System operator**	**Snake Nerve system**

Differences are presented in bold

Table 11.8 Parking reducer case study: ideality strategies comparison

Biological model	Biomimetic concept	
Increase benefits		
More functions	Multifunctional design— The system contains and moves the vehicle. The sleeve protects the vehicle against thieves, vandalism and climate	Multifunctional design— The muscles that lead the prey are also used for the snake locomotion. The prey is contained in the body and kept away of competitors
Reduce costs		
Defensive strategy	The system prevents accidents during parking and saves repairing costs	Reduction of external disturbances —The predators hide the prey against potential competitors
Opportunistic Strategy	Adjustment of structure to function —The contraction of the air pillow is adjusted to the required movement	Adjustment of structure to function —The location of contraction is adjusted to the required movement.
		Usage of available energy sources—The snake energy movement is used also for moving the prey
Prevent waste	Synchronization—The system provide real time data on free parking space, helping to save time, costs and air contamination during the paring search.	Synchronization—Swallowing movement is synchronized with locomotion movement—Saving energy
	Give up redundant parts—The waste of parking space is prevented and parking cost is reduced	

v. Compare biomimetic concept to biological solution (Tables 11.7 and 11.8)

In Table 11.7 we compare the biomimetic and biological system viable models. We see that the functions remain the same in both models while the vehicle is analogical to the prey and the sleeve is analogical to the digestion system. In order to imitate the movement of the prey, there is a need to replace the working unit and the engine, i.e., replacing the muscles with the air pillows that are also characterized with time asymmetric properties. As a result, the field is also changed from biological to mechanical pneumatic field. The snake's nerves system is replaced by the parking reducer computer for the control of the action.

In Table 11.8 we compare the ideality strategies of the biological model and the biomimetic concept. We see that the biomimetic concept uses most of the relevant ideality strategies of the biological model, and even includes some new ones.

(f) Summary

The biomimetic concept is based on the prey swallowing process by the snake. Generally, the prey is analogical to the vehicle, and the snake trunk is analogical to the sleeve of the system.

In both cases there is a need to protect the prey/vehicle against potential hazards (competitors, thieves, vandalisms, weather) while moving it towards a desired direction.

In order to imitate the swallowing process there was a need to replace the following system components:

- Control Unit—The nerve system of the snake is replaced with an automated computer control.
- Engine/working unit—The muscles (time asymmetry) are replaced with the air pillows that are also identified with time asymmetry. While the prey movement is unidirectional, the vehicle movement can goes in both direction—forwards and backwards.
- The biological field that activates the muscles is replaced with the mechanical pneumatic field that activates the air pillows.

The biomimetic concept uses most of the relevant ideality strategies. The opportunistic strategy of using the snake locomotion for moving the prey is irrelevant in the biomimetic case as the whole parking system do not need to move in space, though it may be an innovative idea in cases of occasional events, but probably infeasible.

On the contrary, some ideality strategies are demonstrated by the biomimetic concept and do not appear in the biological model, such as the waste of parking space that is prevented. Generally we may say that we didn't lose any degree of ideality during the transfer stage.

Chapter 12
Lab and Field Experiments

12.1 Introduction

The structural biomimetic design method has two major targets: fostering innovation and sustainability. Elaboration on these aspects is presented in Sects. 1.5 and 1.6.

(1) **Innovation**—Biomimetic innovation is related to the analogical transfer and supported by abstraction methods. Therefore, innovation aspects of the structural biomimetic design method are provided by abstracting biological systems based on patterns identifications and a complete viable system model, i.e., the structural abstraction method.

2) **Sustainability**—Biomimetic sustainability is related to incorporating nature sustainability strategies in designs. Therefore, sustainability aspects of the structural biomimetic design method are provided by the ideality framework (Fig. 6.4) and the ideality patterns table (Table 10.10).

In this part of the study we aim to evaluate the innovation and sustainability aspects of the structural biomimetic design method. For this purpose we performed two different experiments to evaluate innovation and sustainability separately. This separation also resulted in a simpler experimental design.

Each one of the experiments included several sub experiments, lab and/or field experiments. Table 12.1 provides the full view of the experimental process and presents the different sub experiments performed to assess innovation and different stages performed to assess sustainability.

- Lab experiments were performed in an artificial environment of academic exercises. One experiment included class exercises given to an undergraduate 3rd–4th year students and few M.Sc. students in mechanical and industrial engineering product development class. The students used such exercises as part of the teaching approach. The exercises used to assess both the innovation and the sustainability aspects of the structural biomimetic design method. Another experiment was a final project towards M.Sc. degree in mechanical engineering

Y.H. Cohen and Y. Reich, *Biomimetic Design Method for Innovation and Sustainability*, DOI 10.1007/978-3-319-33997-9_12

Table 12.1 Innovation and sustainability sub-experiment and stages

	Sub-experiment 1(a): Lab experiment: class exercise		Sub-experiment 1(b): Lab experiment: final projects		Experiment 1(c): Field experiment: industrial workshop	
	Description	*Assessing*	*Description*	*Assessing*	*Description*	*Assessing*
Exp. 1: assessing innovation	Students were asked to develop a biomimetic design concept based on a biological system by the structural biomimetic design method (from biology to an application)	Innovative criteria of design results	M.Sc. students were asked to develop biomimetic solutions by the structural biomimetic design method: From biology to an application and from a problem to biology	Innovative criteria of design results	Workshop results are biomimetic concepts to address real challenge of the company. The design concepts were developed by the structural biomimetic design method	Innovative criteria of design results
	Stage 1		*Stage 2*		*Stage 3*	
	Description	*Assessing*	*Description*	*Assessing*	*Description*	*Assessing*
Exp. 2: assessing sustainability	Students were asked to identify sustainability aspects with and without different sustainability tools (ideality and life principles)	Comparing the ideality tool to the life principle tool in "sustainability identification" measure	Students were asked to suggest changes to the biological systems so it will incorporate more sustainability principles (ideality and life principles)	Comparing the ideality tool to the life principle tool in "sustainability design" measure	Students were asked to report their perceptions of the sustainability tool (ideality and life principles)	Comparing the ideality tool to the life principle tool in user's perceptions regarding the tool

by two mechanical engineers who devoted serious time in developing biomimetic solutions using the structural biomimetic design method. Part of their solutions was introduced previously as the case studies in Chap. 11. Their design results were used to assess the innovation aspect.
- Field experiment was performed in industry at a medical company and addressed a real challenge. This experiment was used to assess the innovation aspect.

12.2 Experiment 1: Assessing Innovation Aspects of the Structural Biomimetic Design Method

12.2.1 Experiments Rationale

In this part of the experiment we aim to identify the innovation aspects of the structural biomimetic design method. As abstraction is the core of the biomimetic innovation, we focused on this stage. The ideal experiment design could be a classical design experiment comparing the structural biomimetic abstraction method to other abstraction methods. However, due to lack of expertise in other methods we cannot guarantee for the same teaching level as we can for the structural abstraction method. In addition, other abstraction methods differ in terms of complexity and require different time for teaching and practicing. Therefore, we decided to focus on the structural biomimetic method and understand its absolute innovation aspects. We defer to future research a comparison with other biomimetic design methods. Meanwhile, we continue to test the method presented in this book in more elaborate case studies.

12.2.2 Innovation Assessment Process

Nelson et al. [212] referred to evaluation of idea generation techniques and identified two main categories: process-based and outcome based approaches. Process-based approach refers to the cognitive processes that explain the creative thought. Outcome based approach refers to the design outcomes. Process-based approach is more complicated and therefore less prevalent [213]. In this experiment, we used the outcome based approach, evaluating the innovation technique by its design outcomes.

Design outcomes are usually assessed by judges that provide subjective rating [214]. Innovative design concepts are those that are deemed as innovative by domain experts that judge them. Judges may refer to absolute or relative innovation.

Relative innovation is defined when the design concepts are compared to other solutions [212, 215]. In class experiments, design concepts may be compared to other student's designs. In Industrial experiment, design concepts may be compared to other solutions in market. Nelson et al. [212] used this approach in assessing the designs generated by students from a capstone design class and a bio-inspired design class.

Absolute innovation is defined when each concept is directly compared with a set of criteria or categories [212, 215]. For example, an outcome may be categorized according to defined levels of innovation where the highest level represents a paradigm shift. A similar idea of levels of innovation appears at the TRIZ literature [23]. As we mainly aim to understand various innovation aspects of the design outcomes we focus on comparing the design outcomes to a set of innovation criteria (absolute innovation).

12.2.3 Innovation Criteria

Innovation is difficult to characterize in terms of specific features. It is even more complicated when dealing with design concepts that are presented as ideas and not as final products. Several selected efforts to define innovation criteria of design concepts are presented in Table 12.2. Shah et al. [213] identified metrics for measuring ideation effectiveness. Cheong et al. [110] assessed the extraction and transfer of biological analogies for creative concept generation and offered a three

Table 12.2 Innovation criteria for design concepts

	Shah et al. [213]	**Cheong et al.** [110]	**Cropley and Cropley** [216]
1	Novelty - a measure of how unusual or unexpected an idea is as compared to other ideas.	Novelty – "Novel concepts are new, original and surprising."	Novelty - If the concept is relevant and novel, then the product/solution is original.
2	Quality - a measure of the feasibility of an idea and its ability to meet the design specifications.	Usefulness – "Useful concepts are valuable, functional, practical and feasible. They solve the problem."	Relevance & Effectiveness - a concept that solves the problem it is intended to solve. If only relevance is satisfied the concept is routine.
3	Variety - a measure of the solutions space. The generation of similar ideas indicates low variety.	Cohesiveness – "Cohesive concepts appear whole, well developed and are detailed enough to be understandable".	Elegance - If concept is also pleasing to look and goes beyond the mechanical solution, it is elegant.
4	Quantity - the total number of ideas generated. The more ideas, the better the chance to find better ideas.		Generalizability – If the concept is also broadly applicable to different situations and open new perspectives, it is innovative.

component measures for creativity. Cropley and Cropley [216] defined four-dimensional hierarchical model of an innovative creative product. Though this definition was related to products, it can also be used to evaluate concepts.

The shaded cells in Table 12.2 are the similar innovative criteria in all sources. Novelty appears in all three sources. Novel objects are unusual, original, statistically infrequent or unique [217] and evoke "surprise" or "shock" as they are unexpected [218]. However, novelty is not a sufficient factor to assess innovation, as something may be novel but has no utility. Innovation should embody usefulness in order to realize the impact [219]. Shah et al. [213] referred to this aspect by the criteria of 'Quality', an ability to meet the design specifications. Cheong et al. [110] referred to this aspect by the criteria of 'Usefulness', valuable concepts that solve the problem. Cropley and Cropley [216] referred to this aspect by the criteria of 'Relevance', products that solve the products they intended to solve.

Cheong et al. [110] definition of usefulness is rich and may imply several sub-criteria of usefulness including the concepts of value, functionality and feasibility. We elaborated this definition to include also the concept of patentability and offered the following hierarchical sub-criteria to usefulness:

(i) **Value**—First, a concept should aim to solve a valuable problem. If the problem solved is not valuable, the innovation is irrelevant. This sub criterion appears in Cheong et al. definition of usefulness.
(ii) **Functionality**—Second, assuming the problem aimed to be solved is valuable, the concept should indeed solve the problem it aimed to solve. This definition is correlated with Cropley and Cropley [216] definition of 'Relevance', Cheong et al. [110] definition of "solve the problem" and Shah et al. [213] definition of "meet design specification".
(iii) **Feasibility**—Third, assuming that the problem has a value and that the design concept indeed solves that problem, then it should be feasible. If we cannot implement the design concept due to manufacturing or budget limits this innovation is irrelevant. The feasibility aspect appears both in Shah et al. [213] and Cheong et al. [110] definition.
(iv) **Patentability**—Last, assuming the proposed design concept meets the previous criteria, it has more value if it can be also patentable.

In relation to the other assessment criteria of design concepts, offered by Shah et al. [213], variety and quantity are irrelevant to our context as we deal with one concept of each student while variety and quantity refer to a group of concepts. In relation to the other assessment criteria of design concepts, offered by Cheong et al. [110], we found that cohesiveness is an applicable and valuable criteria in our context as we would like to assess in what manner the structural biomimetic abstraction method can create complete and well developed design concepts, assuming that the more complete and detailed design concepts, the better chance they have to end up with a real innovation. In relation to the other assessment criteria to design concepts, offered by Cropley and Cropley [216] we found that both elegance and generalizability may be difficult to assess in our context, due to

Table 12.3 Experiment rubric: innovations criteria

	Criteria	Sub-criteria
1	Novelty	**Uniqueness**—The design concept is original (unusual)
		Surprise—The design concept is surprising (unexpected)
2	Usefulness	**Value**—The design concept solves a valuable problem. (If the problem solved is not valuable, the novelty is irrelevant)
		Functionality—the design concept indeed solves the problem it aimed to solve. (Achieves its declared function)
		Feasibility—The design concept is feasible in a reasonable cost-effective ratio. (If we can't perform the design concept due to manufacturing or budget limits its novelty aspects are irrelevant)
		Patentability—The design concept may be patentable
3	Cohesiveness	**Cohesiveness**—The design concept appears complete, well developed and it is detailed enough to be understandable [110]
4	General innovation	**General innovation**—The design concept is innovative according to my experience
		Breakthrough innovation—The design concept is a breakthrough innovation offering a paradigm shift
5	Relative innovation	**Relative innovation**—The design concept is innovative relative to other designs/solutions in market

the limited time that is devoted to develop the concept by the experiment participants. Originally, these criteria were offered to assess products and not concepts. As a result, we offer the following innovation criteria rubric presented in Table 12.3 to assess the innovation aspects of the structural biomimetic design outcomes. Though different weights may be assigned to each criterion, as offered by Oman et al. [219] in their Comparative Creativity Assessment measure, in our study, each criterion will get the same weight.

12.2.4 *Experiment 1(a)—Lab Experiment in Class*

Industrial and mechanical students in a product development class got a general explanation about bio-inspired design including the main design stages from biology to an application. Following this explanation, the students were exposed to the structural abstraction method. Exposure included explanation and three examples of biological system abstractions. These explanations were given by the authors.

Two different biological systems at the same level of complexity were distributed randomly among the students of each group. The biological systems are presented in Appendices A and B. Students were asked to develop a biomimetic design concept based on the biological system. The biological system served as a stimulus to assist the creative biomimetic design process [220]. Each one of the stimulus included enough information so superficiality will not inhibit the ideation process [221].

Table 12.4 Innovation experiment design

Group	Stage 1	Stage 2	Stage 3
Experiment group	General explanation about bio-inspired design	Studying the structural abstraction method	Solving problems—Develop a biomimetic design concept
Time in minutes	25	50	30

Experiment design is presented in Table 12.4. Instructions and forms used in this experiment appear in Appendices C–E.

Assessment Procedure Two innovation experts assessed each design concept by Likert scales (1–10), ranking each one of the innovation criteria presented in Table 12.3. The assessment form appears in Appendix E. We didn't use the VAS scales (Visual Analogue Scale) though it is a continuum scale adjusted for subjective characteristics, due to operational constraints. Relative innovation was not assessed as we did not find its value in the context of "lab" experiment. We compared the average rate of each innovation criteria.

Results 11 students, 3 Industrial and 8 mechanical engineering students, participated in this study as part of an engineering design course. Most of them were B.Sc. students in their last year of studies and few were M.Sc. students. 8 of them were males and 3 females. All students had relatively low levels of knowledge in design (2.3 in average) and in biomimicry (1.4 in average), according to their reports in scale of 1–10. In Table 12.5 we present the innovation experts assessments according to the experiment rubric.

Discussion Two judges rated the innovation criteria of the design outcomes of the structural abstraction method. The reliability between the judges is relatively high (correlation coefficient of 0.75). Design outcomes are derived from two different biological mechanisms, thus we control, at least partially, the type of stimulus (biological system) effect. Generally the ratings are low with a total average of 2.8 based on 1–10 scales. However, considering the relatively short time devoted for the development of these biomimetic concepts, some of the low rating are reasonable. For example, it is not expected to yield a patentable or breakthrough innovation in a lab experiment taking less the 30 min.

We can infer about the distribution of ratings between the innovation criteria. Thus for example, the functionality and feasibility got the highest rates (4.13 and 4.22 accordingly) and the cohesiveness and value yielded medium rating (3 and 3.13 accordingly). The structural abstraction method accesses a complete functional system model of the biological system to engineers, enriched with details about the structure-function relations. So it is reasonable to realize that the value, functionality, feasibility and cohesiveness aspects are relatively high, as these aspects are related to functionality.

Table 12.5 Experiment 1(a): expert's innovation assessment

	Uniqueness	Surprise	Value	Functionality	Feasibility	Patentability	Cohesiveness	General innovation	Breakthrough
Expert 1	2.36	2.18	2.9	4.9	4.81	1.9	3.09	1.81	1.19
Expert 2	3.09	3	3.36	3.36	3.63	2.54	2.9	2.63	1.09
Avg (2.8)	2.725	2.59	3.13	4.13	4.22	2.22	3	2.22	1.14
Stdv	0.365	0.41	0.23	0.77	0.59	0.32	0.095	0.41	0.05

Novelty criteria (uniqueness and surprise) yielded relatively low rates. An explanation for that may be the fact that due to the time limitation, each student had only one biological system as a source of inspiration. Thus, the innovation is limited to solutions that are related to this biological system. We assume that by introducing more biological systems or by performing "from a problem to biology design process" and searching for several biological systems, novelty aspects could be higher.

The major disadvantages of this lab experiment are the limited time devoted for developing the biomimetic design concept (about 30 min), limited scope, reflecting only the abstraction part of the structural biomimetic design process, and one design direction, from biology to an application. Therefore the conclusions are limited.

12.2.5 Experiment 1(b)—Lab Experiment as Final Project

Following the major disadvantages of the previous lab experiment, limited time, scope, and developing direction, we repeated the innovation assessment with biomimetic concepts developed during a period of few weeks and by using the complete structural biomimetic design algorithm, presented in Sect. 10.1. Two mechanical engineers studied the structural biomimetic design method from the authors and developed 6 biomimetic design concepts, submitted as final projects of their M.Sc. degree in mechanical engineering. Four concepts were developed from biology to an application and two from a problem to biology. Some of them were presented in Chap. 11. We assessed the six biomimetic design concepts.

Assessment Procedure and Results Two innovation experts, the same ones of experiment 1(a), assessed these biomimetic designs by Likert scales (1–10), ranking per the innovation criteria rubric (Table 12.3). The assessment form appears in Appendix E. We did not use the VAS scales (Visual Analogue Scale) though it is a continuum scale adjusted for subjective characteristics, due to operational constraints. Relative innovation was not assessed as we didn't find its value in the context of "lab" experiment. Their ratings are presented in Table 12.6. In addition, the two mechanical engineers who developed the biomimetic concepts rated their own design outcomes according to the same innovation criteria by Likert scales (1–10). Their ratings are presented in Table 12.7.

Discussion Two judges rated the innovation criteria of the design outcomes of long term biomimetic projects, developed by the structural biomimetic design method. The reliability between the judges is lower in this case (correlation coefficient of 0.57), as one of the experts systematically rated higher most of the innovation criteria.

Generally the ratings are higher compared to experiment 1(a) with a total average of 5.23 based on 1–10 scales comparing to 2.8 in experiment 1(a). Rates are higher this time probably due to the extended time and effort devoted in these

Table 12.6 Experiment 1(b): expert's innovation assessment

			Uniqueness	Surprise	Value	Functionality	Feasibility	Patentability	Cohesiveness	General innovation	Breakthrough
From a problem to biology	Parking reducer	Expert 1	7	7	7	7	6	7	6	5	1
		Expert 2	3	2	2	2	3		2	3	1
		Avg.	5	4.5	4.5	4.5	4.5	7	4	4	1
	Dynamic screen cover	Expert 1	7	7	6	8	7	6	6	5	1
		Expert 2	3	2	2	2	3		2	5	1
		Avg.	5	4.5	4	5	5	6	4	5	1
From biology to an application	Cooking pot with temp. control	Expert 1	7	7	8	8	7	8	8	6	1
		Expert 2	3	3	3	5	8		8	2	1
		Avg.	5	5	5.5	6.5	7.5	8	8	4	1
	Color changing sheet	Expert 1	7	7	8	8	7	8	7	6	2
		Expert 2	4	5	3	5	5		5	4	2
		Avg.	5.5	6	5.5	6.5	6	8	6	5	2
	Unidirectional humid valve	Expert 1	6	6	6	7	7	5	7	5	1
		Expert 2	5	7	5	6	6		7	4	4
		Avg.	5.5	6.5	5.5	6.5	6.5	5	7	4.5	2.5
	Leading and releasing plug	Expert 1	7	7	7	8	7	7	7	6	2
		Expert 2	5	7	5	6	6		5	5	4
		Avg.	6	7	6	7	6.5	7	6	5.5	3
	Total Avg.	5.23	5.3	5.6	5.2	6	6	6.7	5.8	4.7	1.8

Table 12.7 Experiment 1(b): designer's innovation assessment

		Uniqueness	Surprise	Value	Functionality	Feasibility	Patentability	Cohesiveness	General innovation	Breakthrough
From a problem to biology	Parking reducer	10	9	10	10	5	7	5	9	7
	Dynamic screen cover	7	6	4	5	5	4	6	6	2
	Avg.	8.5	7.5	5	5.5	5	5.5	5.5	7.5	4.5
From biology to an application	Cooking pot with temp. control	8	7	7	8	7	7	6	9	6
	Color changing sheet	6	6	6	9	6	7	6	8	5
	Unidirectional humid valve	10	10	8	9	7	8	7	10	9
	Leading and releasing plug	9	7	7	8	6	5	5	8	6
		8.25	7.5	7	8.5	6.5	6.75	6	8.75	6.5
Total Avg.	7.0	8.3	7.5	7.0	8.2	6.0	6.3	5.8	8.3	5.8

projects, using the full structural biomimetic design method this time, and not only its abstraction part.

Observing the distribution of ratings between the innovation criteria reveals similar results to part 1(a). The functionality and feasibility aspects yielded relatively high rates around 6 in avg. while the cohesiveness is following with 5.8 in Avg. The structural biomimetic design method is function oriented, providing a platform to model biological systems with details to support the functionality, feasibility and cohesiveness aspects of the design concept. Novelty criteria (uniqueness and surprise) yielded relatively higher rates this time compared to experiment 1(a), probably due to the extended effort.

The designers themselves were satisfied with the innovation aspects of their designs, rating the general innovation relatively high (8.3 out of 10). The designers appreciated the novelty aspects of their designs rating them high in uniqueness and surprise (8.3 and 7.5 respectively). Again, functionality got the highest rate, as the designers believed that their designs indeed achieve their declared function. The sample is not large enough to compare design directions (from biology to an application vs. from a problem to biology).

12.2.6 Experiment 1(c)—Field Experiment in Industry

The major disadvantage of the previous experiments is the fact that they are limited "lab" experiments, performed in an artificial environment of academic exercise. In this part we wanted to assess the structural biomimetic design method in a real challenge.

For this reason we performed a biomimetic workshop at a medical company, aimed to address a real design challenge by the structural biomimetic design method. Due to confidentially aspects, we describe the process and results without detailing the specific developed solutions.

The Company The company develops and manufactures cerebral protection devices that aim to reduce the risk of stroke associated with cardiovascular procedures. The company has a developed solution. The target of the workshop was to observe other biomimetic potential solutions (Remark: the author who managed this workshop was not exposed to the company current solution until the end of the workshop).

The Framework The biomimetic design concepts were developed during four meetings in sequential weeks. The workshop was given by the author. Ten persons of the senior management of the company attended the meetings including the CEO and CTO. Each meeting took 3 h. The meetings contents and results are presented in Table 12.8.

Between the second and the third meeting both participants and the author devoted time for further search of relevant biological systems and abstracting the solutions per the structural patterns.

Table 12.8 Experiment 1(c): industrial experiment meetings

	Subjects	Results
Meeting 1	a. Introduction to biomimetic design b. Introduction to the structural biomimetic design method	
Meeting 2	a. Analyzing the design space b. Organism search	a. Function-means tree—different design paths to achieve the design challenge b. Relevant organisms to the identified design paths
Meeting 3	a. Abstraction of biological solutions by structure-function patterns b. Introduction of possible biomimetic design concepts by the three basic parts of the biomimetic innovation: Biological system-abstraction-transfer	a. Six biomimetic design concepts. Each one represents a solution for different design path
Meeting 4	Open Discussion on each one of the six biomimetic solutions. Enriching and further developing of the most promising concepts	a. Two biomimetic design concepts worthwhile further evaluation

Assessment Procedure and Results Each one of the two final biomimetic concepts was assessed by the innovation criteria rubric (Table 12.3) with 1–10 VAS scales (Visual Analogue Scale), as this response scale is adjusted to subjective characteristics. The value criterion was removed as it is not a distinctive criterion as all the concepts referred to the main design challenge of the company and the cause of its establishment. We assume that this challenge has a value if the company exists. Relative innovation was assessed by comparing the workshop design results to other solutions in the market.

The judge was the CTO of the company who has over 20 years of experience in innovative design and is highly familiar with the domain solutions and patents. His ratings are presented in Table 12.9. In additions we interviewed the CTO and gained some qualitative assessment regarding the process in general and its innovation aspects in particular. The evaluation and interview took place 6 months after the workshop.

Six months after the workshop, the author interviewed the CTO to get an update about continuous evaluation and promotion of the workshop final design concepts. Concept 2 was not further developed due to manufacturing constraints. Though this concept yielded relatively high innovation rates, the company team didn't know how to proceed with its manufacturing. However, the CTO reported that in the presence of new manufacturing technologies, this concept may be further developed. Concept 1 was not further developed as it was not unique and resembled in some aspect current solution in market.

Generally, the CTO reported that the workshop created a positive vibe in the company, resulted in further discussions and activities beyond the workshop time frame. The team was inspired by the process. The CTO assumed that this

Table 12.9 Experiment 1(c): expert's innovation assessment

	Uniqueness	Surprise	Functionality	Feasibility	Patentability	Cohesiveness	General innovation	Breakthrough	Relative Innovation
Biomimetic concept 1	2.7	7.7	4.7	8	5	7.3	7.4	1.2	9.3
Biomimetic concept 2	8.4	8.3	6.3	5	8.5	8.5	8.3	5.6	9
Total Avg.	5.55	8	5.5	6.5	6.75	7.9	7.85	3.4	9.15

innovation process has much to contribute especially to homogeneous teams who are not used to interdisciplinary design processes. The CTO could not compare the workshop to other innovation tools as he was not familiar with other tools.

Discussion The workshops yielded innovative concepts also when compared to market solutions. (9.15 in avg.) The general innovation is high (7.85 in avg.). The cohesiveness aspect is relatively high (7.9 in avg.), and the functionality, feasibility, and patentability got medium rates (5.5–6.75 in avg.). The structural biomimetic design method accesses a complete functional system model of the biological system to engineers, enriched with details about the structure-function relations. So it is reasonable to realize that these aspects are relatively high. In contrast to the lab experiments, the novelty (uniqueness and surprise) got higher rates (5.5 and 8 in avg.). When starting with engineering challenge and devoting enough time for the organism search and concepts development we can see also the novelty aspects of this design method.

12.2.7 Summary: Experiment 1—Assessing Innovation Aspects of the Structural Biomimetic Design Method

We aimed to identify the innovation aspects of the structural biomimetic design method and used three sub experiments for this purpose. Each experiment elaborated the scope of innovation assessment by addressing new aspects. We started with assessing the innovation criteria of a class experiment that was based on a relatively short exercise. We moved to assess the innovation criteria of final projects that required longer time. Finally, we went out of lab and moved to assess a real case in industry. Thus, the devoted time and the location of experiment is changing and increasing the scope of data. Indeed we observed higher rates when we moved from class to a final project and from final project to industry, mainly at the expert's evaluations. Innovation criteria assessment is presented per users and per experts in Tables 12.10 and 12.11 respectively. The assessments of the CTO in relation to the industrial experiment appear in both tables, as he serves both a user and as a domain expert. Although these tables average data obtained in different ways, they summarize all the experiments data in one place and in a new perspective of experts versus users evaluation.

The general innovation criteria of the structural biomimetic design method yielded high rates by users (8.07). The innovation is high even when compared with other solutions in market (9.15), however the criteria of breakthrough innovation got only medium rate (4.6). Experts tended to evaluate the general innovation as high when more effort was devoted in the concepts development (experiments c and b vs. a) but generally evaluated the criteria of breakthrough innovation as low (2.1).

Medium to high innovation rates were given by users (6.25–6.85) and by experts (5.2–5.6) to criteria related to translating the design concept into reality: functionality, feasibility, patentability and cohesiveness. The structural biomimetic

Table 12.10 Innovation experiment summary: user's evaluations

	Uniqueness	Surprise	Value	Functionality	Feasibility	Patentability	Cohesiveness	General innovation	Breakthrough	Relative innovation
1(b)—Designers	8.3	7.5	7	8.2	6	6.3	5.8	8.3	5.8	
1(c)—CTO as user	5.5	8		5.5	6.5	6.75	7.9	7.85	3.4	9.15
Avg.	6.9	7.75	7	6.85	6.25	6.52	6.85	8.07	4.6	9.15

Table 12.11 Innovation experiment summary: expert's evaluations

	Uniqueness	Value	Surprise	Functionality	Feasibility	Patentability	Cohesiveness	General innovation	Breakthrough	Relative Innovation
1(a)—Experts	2.7	2.6	3.1	4.1	4.2	2.2	3.0	2.2	1.1	
1(b)—Experts	5.3	5.6	5.2	6.0	6.0	6.7	5.8	4.7	1.8	
1(c)—CTO as domain expert	5.5	8		5.5	6.5	6.75	7.9	7.85	3.4	9.15
Avg.	4.5	5.4	4.2	5.2	5.6	5.2	5.6	4.9	2.1	9.15

design method is functional oriented, providing a platform to model functional aspects of biological systems with cohesive details to support its realization. Novelty criteria including uniqueness and surprise yielded high rates by users (6.9–7.75) and medium rates by experts (4.5–5.4). The more effort is devoted to develop the concepts, the higher are their novelty and value rates.

We acknowledge that other innovative design concepts could be achieved using different biomimetic design approaches. The experiments in this section describe the innovation aspects of the structural biomimetic design method. Future research should compare them to innovation aspects of other biomimetic design methods.

12.3 Experiment 2: Assessing the Ideality Framework as a Sustainability Analysis and Design Method

12.3.1 Experiment Rational

Sustainability tools aim to assess sustainability levels of a given system and implement sustainability during the design development process. Elaboration about sustainability tools may be found in Sect. 1.6.

In biomimetic design, nature sustainability design principles are identified and implemented during the design process. There are two major activities being done with these tools in relation to biomimetic design: (i) Identification—Identifying sustainability principles at a given biological system. (ii) Implementation—Suggesting a new design/design change based on these principles.

In this study, we compare two biomimetic sustainability tools: the life principles (Sect. 1.6.3.1) and the ideality (Sect. 9.3) tools. Each tool provides a framework to identify sustainability principles, some similar and some different. We decided to choose the life principles as a base for comparison as it is the only sustainability tool that is derived from the biomimetic literature. Life principles are nature sustainability design strategies and in this manner they resemble the ideality strategies. They also may be used during early design stages and are simple enough to be thought under the time frame of the experiment. However, life principle knowledge base is extended and includes 6 main strategies and 20 sub strategies or design principles. Due to time limitation and in order to balance between the two tools, we chose only three strategies out of the six as a base for comparison, the ones that most resemble the ideality strategies. ('Adapt to changing conditions', 'Be locally attuned and responsive', 'Be resource efficient').

First we want to assess if there is a difference in various sustainability measurers between a state of using sustainability tools (ideality and life principles) and a state of using no sustainability tools. In addition, as some of the life principles are general and their application in engineering is neither clear nor straightforward, we want to assess if there is a difference between the life principles and the ideality tool.

12.3.2 Measurers

Following the experiment rationale, we defined the following measures to assess the tools effect:

Sustainability Principles Identification This measure deals with the student's ability to use the tool for identifying sustainability principles of a given biological system. We assess this measure by the following sub-measures:

(i) Tool Validity—Do students who use the sustainability tools indeed identify the sustainability principles that can be extracted by these tools?
(ii) Tool Reliability—Whether different students who use the tools achieve similar results?

Sustainable Design This measure deals with the student's ability to use the tool for sustainable design activities. We assess this measure by the following sub-measures:

(i) Number of changes suggested—How many design changes are offered to enhance system sustainability, by using each one of the tools?
(ii) Perceived difficulty to use the tool for design activities—reported by students.

Perceived Usability This measure deals with the student's personal experience in using the tool. We assess this measure by the following sub-measures regarding to each one of the tools:

(i) Perceived difficulty to use the tool for identifying sustainability principles*.
(ii) Level of sustainability understanding*.
(iii) Satisfaction—participant satisfaction from the sustainability analysis*.
(iv) Future use—Likelihood to use this tool in the future.

*These measures are assessed before and after using the tools and are presented as change variables, the difference between the measures after using the tool minus the measure before using the tool.

12.3.3 Experiment Hypotheses

1. Students who are exposed to the sustainability tools (ideality or life principles) will get higher scores in each one of the change variables comparing to state of using no sustainability tools.
2. Students who are exposed to the ideality tool will get higher scores in the sub-measures of **Sustainability principles identification** comparing with students who are exposed to the life-principles tool.
3. Students who are exposed to the ideality tool will get higher scores in the sub-measures of **Sustainability design** comparing students who are exposed to the life-principles tool.

4. Students who are exposed to the ideality tool will get higher scores in the sub-measures of **Perceived usability** comparing with students who are exposed to the life-principles tool.

12.3.4 Experiment Design

The research hypotheses were validated by a classical experiment design, including an experiment and a control group. This experiment is a lab experiment and was conducted at a class of industrial and mechanical engineering student in a product development course. Experiment design is presented in Table 12.12.

All students were first allocated randomly into one of two groups:

- Group A: Experiment group—exposure to the Ideality tool.
- Group B: Control group—exposure to "Life principles" tool.

Stage 1: Before Exposure to Sustainability Tools All students (experiment and control groups) got a general explanation about sustainability including the definition of sustainability. No sustainability tools were given at this stage. This stage was given by an instructor who is a domain expert in sustainability.

Stage 2: Sustainability Analysis—Problem 1 Two different biological mechanisms at the same level of complexity were distributed randomly among the students of each group. Hence, we control the effect of the system type that might be a confounding variable. Students were asked to identify the sustainability principles of the given biological system. Biological systems for analysis are presented in Appendices F and G. Student's instructions are presented in Appendix H.

Stage 3: Exposure to Sustainability Tools Each group was exposed separately to a sustainability tool. The experiment group was exposed to the Ideality tool. Exposure included explanation and three examples of biological system ideality analysis. The control group was exposed to the "life principles" tool. Exposure

Table 12.12 Sustainability experiment design

Group	Stage 1	Stage 2	Stage 3	Stage 4
Experiment group	General explanation about sustainability	Analysing biological system A1	Studying the "Ideality" tool	Analysing biological system A2
Control group		Analysing biological system B1	Studying "Life principles" tool	Analysing biological system B2
Time in minutes	20	20	45	20

included explanation and three examples of biological system life principles analysis (same biological systems as the experiment group but analyzed with different sustainability tool). The experiment group session was given by the author and the control group was given by another instructor who is a domain expert in sustainability and highly familiar and experienced with the life principles tool. The same time was devoted to the exposure at this stage.

Stage 4: After Exposure to the Sustainability Tools At this stage each student analyzed the biological systems he did not analyze in the first stage. Students were asked to identify the sustainability principles of the given biological system by using the sustainability tool presented to them (Ideality or life principles). In addition, students were asked to suggest as many design changes as possible to these biological systems in order to increase their sustainability. Student's instructions for experiment and control groups are presented in Appendices I and J accordingly. In addition to these instructions every group got as a reference a list of ideality/life principles including examples.

Assessed Effects The ideality analysis method effect is assessed by comparing the results of A1 and A2 (before and after the exposure to the ideality method) and by comparing A2 to B2 (After exposure to different sustainability tool—the life principles).

Time/learning effect may not be assessed by this experiment design. It may be assumed that as both groups have the chance to learn from the first sustainability analysis, if we look at the difference between A2 and B2 it represents the sustainability tool effect.

12.3.5 Measuring Process

12.3.5.1 Sustainability Principles Identification

Experts in "Ideality" and "Life-principles", who have experience in using these tools for sustainability analysis and design, identified the sustainability principles of each one of the biological systems by using their tool of expertise. One of the experts is the author who analyzed the biological systems with both tools. The life principle other expert is also the instructor of this session. The Ideality other expert is a TRIZ specialist who is highly experience in teaching ideality in academic courses.

Each biological system was analyzed by the two different experts separately, and their solutions were integrated to create one "expert's solution" for each biological system. We assume that these expert's solutions represent the sustainability principles space as may be inferred by each one of the sustainability tools. The expert's solutions are presented in Appendices K–N.

We used these expert's solutions to assess the validity and reliability of the measure "Sustainability principles identification".

(i) "Sustainability principles identification"—Validity

Definition Level of resemblance between the student's sustainability analyses to expert's sustainability analyses. This comparison assesses if the tools indeed provide all the sustainability principles that can be extracted by this tools, as defined by the experts.

Assessment Procedure The expert's solutions were translated into a list of '1' and '0' as presented in Appendices K–N. Each ideality/life principle identified at the given biological system was ranked with '1', and otherwise with '0'. The result is a list of '1' and '0' that represents the sustainability analysis by experts for each biological system per each sustainability tool. Same ranking was done for each student analysis. We compared each student analysis list of '1' and '0' to the expert solution list of '1' and '0' and calculated the following measures:

- Total accuracy—percentage of student's correct identification of sustainability principles (both students and experts identified).
- Total Miss—percentage of principles that the student did not identified while the experts did.
- Total Wrong—percentage of principles that the student identified as exist while the experts did not.

The higher the accuracy and the lower the total miss and wrong, the higher is the validity of the tool as it provided better identification of sustainability analysis.

(ii) "Sustainability principles identification"—Reliability

Definition Level of resemblance between the students sustainability analyses to other student's sustainability analyses. This comparison assesses if the tools are reliable and provide similar results by different users.

Assessment Procedure We compared the sustainability analysis lists of '1' and '0' of each student to the other student's lists using the same tool (ideality/life principle).

For each sustainability principle we calculated the percentage of similarity between the students answers, i.e., how many students provided the same identification, '1' or '0', even though this identification was not right according to the expert's solution (repetition 1). We repeated the reliability assessment but now counted only repetition on right identifications, same identification as the experts (repetition 2).

The higher the percentage of similarity between the student's lists using the same tool, the higher is the reliability of the tool.

12.3.5.2 Sustainable design

Definition The ability to use the tool for design activities such as suggesting a design change, and not only for identifying principles. We assess this measure by the following sub-measurers regarding each one of the tools:

(i) Number of changes suggested

Definition How many design changes did the student offer to a given biological system in order to improve its sustainability? (Based on the tools).

Assessment Procedure We coded each different sustainable design change offered by each student to a given biological system in order to improve its sustainability and counted how many sustainability changes were offered after stage 4 in each one of the tools. At this stage we compared the number of changes between the tools.

As the life principles tool provides 12 principles while the ideality tool provides only 10, the probability to suggest more changes in the life principle tool is increased. In order to fix this distortion we reduced the changes suggested by the life principle tool in the ratio of 10/12, and referred to the fixed data as a base of comparison between the tools.

(ii) Difficulty to suggest a change

Definition How difficult was the process of suggesting changes to the system in order to enlarge its sustainability?

Assessment Procedure We assessed this measure by student's response scales (Appendix P, question #4). We used the VAS (Visual Analogue Scale) response scales as it is adjusted to subjective characteristics.

12.3.5.3 Perceived Usability

Each one of the sub-measurers was assessed by student's response scales. We used the VAS (Visual Analogue Scale) response scales as it is adjusted to subjective characteristics. The students were required to specify their level of agreement to statements by locating a position along a continuous line between two end-points. VAS response scale is a continuous and not discrete scale, therefore appropriate for a wider range of statistical analysis including the t-tests.

Student's questionnaires after stage 2 and 4 appear in Appendices O and P responsively.

12.3.5.4 Confounding Variables

- Confounding variables that are controlled by allocating students randomly to one of the groups: gender, age, previous experience in sustainable design, and previous knowledge in sustainability.
- Type of system effect is controlled by the fact that half of the students started with one biological system and the other half with the other.

12.3.6 Statistical Analysis—Results

Subjects
38 students, Industrial and mechanical engineering students, participated in this study as part of an engineering design course. Most of them were B.Sc. students in their last year of studies and few were M.Sc. students. Students were allocated randomly to one of the groups. There were no major differences between the groups in terms of age, gender, academic backgrounds, and previous knowledge in sustainability and sustainable design, as presented in Table 12.13. Hence, the possible confounding variables are controlled.

Hypothesis 1 Students who are exposed to the sustainability tools (ideality or life principles) will get higher scores in each one of the change variables compared with a state of using no sustainability tools.

Several measures, presented in Table 12.14, were measured before and after using the tools and presented as change variables (rate after using the tool minus the rate before using the tool). We performed a Signed Rank Test (Mu0 = 0) that checks if their average is ≥ 0 to assess if the change is indeed significant. All the change variables are significantly ≥ 0 in the two tools, ideality and the life principles.

Table 12.13 Sustainability experiment: students profile per group

	Experiment group	Control group	Total
Number of students	19	19	38
Avg. age	26.4	27	26.7
Male	10	12	22
Female	9	7	16
Mechanical engineering	9	11	20
Industrial engineering	10	8	18
Previous knowledge in sustainable design (ranked as 1–10)	0.76	0.5	0.63
Previous knowledge in sustainability (ranked as 1–10)	1.32	1.2	1.26

Table 12.14 Signed rank test to assess Hypothesis 1: change variables

		N	Mean	Std	Min	Max	Median	*Signed rank test (Mu0 = 0)
It was easy to identify the sustainability principles of the biological system (change variable: After-Before)	Ideality	16	1.79	4.17	−7.9	6.7	3.35	P = 0.0225
	Life principles	17	2.86	3.96	−9.1	9.3	3.8	p = 0.0058
I felt confidence during the analysis system (change variable: After-Before)	Ideality	16	2.77	3.15	−3.3	7.1	3.3	P = 0.0053
	Life principles	17	2.05	3.77	−9.2	7.3	2.7	p = 0.0129
I completely understood the sustainability principles of the biological system (change variable: After-Before)	Ideality	16	2.46	2.76	−2.8	6.1	2.25	P = 0.0054
	Life principles	17	1.78	3.6	−9.6	5.5	2.6	P = 0.0136
I am pleased from my sustainability analysis results (change variable: After-Before)	Ideality	16	1.83	2.85	−3.1	7.1	1.75	p = 0.0224
	Life principles	17	1.55	3.66	−9.7	7	1.7	P = 0.0124
Number of sustainability principles identified (change variable: After-Before)	Ideality	16	4.94	1.65	2	8	5	P < 0.0001
	Life principles	17	4.53	1.66	0	7	5	P < 0.0001

After using the tools, students reported on higher levels of confidence, sustainability understanding and satisfaction from the sustainability analysis, as well as less difficulties in the analysis process. Students identified 4.94 and 4.53 more sustainability principles when using the ideality and life principles tools, respectively, in relation to a state of no sustainability tools. The major sustainability principles identified without tools were related to: (i) Survival aspects (ii) Resource efficiency (iii) Not damaging the environment. The sustainability principles identified after using the tools were more diverse and extracted from the variety of possibilities suggested by each one of the tools.

All the change variables, such as confidence, difficulties, sustainability understanding and satisfaction from the sustainability analysis are significantly higher after using the sustainability tools. Consequently, Hypothesis 1 is accepted.

Hypothesis 2 Students who are exposed to the ideality tool will get higher scores in the sub-measures of **Sustainability principles identification** compared with students who are exposed to the life-principles tool.

We performed a two sample t-test (Table 12.15) to assess whether there is a significant difference between the ideality and life principle tools in the

Table 12.15 Two sample t-test to assess Hypothesis 2: sustainability principles identification

	Sustainability principles identification sub-measurers		N	Mean	Std	Min	Max	Median	Two sample t-test
Validity	% Accuracy in identifying sustainability criteria that the experts identified	Ideality	19	77.89	14.75	50	100	80	P = 0.0086
		Life principles	19	64.91	14.05	41.67	91.67	58.33	
	% Miss—Sustainability criteria that the students did not identify and the experts did	Ideality	19	14.74	11.72	0	40	10	P = 0.0005
		Life principles	19	28.95	11.24	8.33	41.7	33.3	
	% Wrong—Sustainability criteria that the students identified and the experts did not	Ideality	19	7.37	8.06	0	30	10	ns*
		Life principles	19	6.14	6.72	0	25	8.33	
Reliability	% Repetition 1—with wrong identifications	Ideality	10	79.4	16.2	52.6	100	81.7	ns*
		Life principles	12	76.7	12.97	52.6	100	78.9	
	% Repetition 2—with right identifications	Ideality	10	77.9	18.35	52.63	100	81.58	ns*
		Life principles	12	64.9	22.56	21.05	100	71.02	

"Sustainability principles identification" sub-measurers including the validity and reliability.

"Sustainability principles identification" **validity** refers to the user's ability to extract "sustainability principles" with this tool. It was assessed by the sub-measures of "*accuracy*", "*miss*" and "*wrong*" presented in Table 12.15. There is a significant difference between the total accuracy of students who used the ideality tool compared to students who used the life principle tool (P = 0.0086). The total accuracy of the ideality group was 77.89 in average while the total accuracy of the life principle group was 64.91 in average. Students who used the life principle tool significantly missed more sustainability criteria identified by the experts compared to students who used the ideality tool (28.95 % miss vs. 14.74 % in average, p = 0.0005). There was no significant difference between the two groups in the measure of total wrong, sustainability criteria that the students wrongly identify and the experts did not.

"Sustainability principles identification" **reliability** refers to the reliability among users, i.e., the similarity of sustainability identifications between different users. It was assessed by counting how many students repeated the same identification, wrong or right ones. Repetition 1 (with wrong identification) of the ideality and life principles tools in average is 79.4 and 76.4 respectively (ns). Repetition 2

Table 12.16 % Repetition between students: ideality and life principles

	Ideality	% Repetition 1	% Repetition 2	Life principles	% Repetition 1	% Repetition 2
1	Multifunctional design	89.47	89.47	Use multifunctional design	78.95	78.92
2	Stronger Effect—Intensify function values	94.74	94.74	Using available energy sources	**68.42**	**42.11**
3	Prevent external disturbances	94.74	94.74	Recycle all materials	**78.95**	**21.05**
4	Reduction of surface area when it causes damage	94.74	94.74	Fit form to function	100.00	100.00
5	Usage of gradients	52.63	52.63	Self—Renewal	**89.43**	**47.33**
6	Adjustment of structure to function	100.00	100.00	Incorporate Diversity	73.65	73.65
7	Transferring functions to the Supersystem	73.68	73.68	Redundancy	89.47	89.47
8	Synchronizing system parameters	**63.16**	**47.37**	Decentralization	57.86	42.11
9	Improving the conductivity of energy through	63.16	63.16	Use readily available Local sources	52.61	52.61
10	Give up redundant parts	68.42	68.42	Use feedback loops	78.92	78.92
11				Leverage cyclic processes	68.38	68.38
12				Cultivate cooperative relationship	84.17	84.17

(with right identifications) of the ideality and life principles tools in average is 77.9 and 64.9 respectively (ns). We see that reliability of the ideality tool is higher in both cases of including or excluding wrong identifications; however, the sample is too small to indicate significance.

The sustainability principles repetition for ideality principles and life principles is presented in Table 12.16. Ideality principles that yielded around 90 % repetition are multifunctional design, stronger effect, prevent disturbances, reduce surface area, and adjustment of structure to function. Synchronizing system parameters is the principle that most of the students missed its identification. Life principles that yielded around 90 % of repetition are 'redundancy' and 'Fits form to function'. 'Using available energy sources', 'Recycle materials' and 'Self-renewal' are the principles that most of the students missed their identification.

Hypothesis 2 is accepted partially. The part of validity is supported statistically; the part of reliability is not supported statistically, but the tendency implies an advantage to the ideality tool that might be significant in larger samples.

Table 12.17 Two sample t-test to assess Hypothesis 3: sustainability design

		N	Mean	Std	Min	Max	Median	Two sample t-test
Number of changes suggested	Ideality	19	2.05	1.84	0	8	2	ns*
	Life principles	19	1.47	1.26	0	4	2	
Number of changes suggested (fixed)	Ideality	19	2.05	1.84	0	8	2	Border Line significance P < 0.01*
	Life principles	19	1.22	1.02	0	3.33	1.66	
Difficulty to suggest changes	Ideality	19	6.60	3.2	0.4	10	6.2	ns*
	Life principles	18	7.06	2.9	0.9	10	8.45	

Hypothesis 3 Students who are exposed to the ideality tool will get higher scores in the sub-measures of **Sustainability design** compared to students who are exposed to the life-principles tool.

We performed two sample t-test procedure (Table 12.17) to assess the differences between the ideality and life principle tools in the "Sustainability design" measurers: number of changes suggested and difficulty to suggest a change. The number of changes suggested by the Ideality tool and the life principles tools was 2.05 and 1.47, respectively (ns). When we repeated the comparison with fixed data as explained in Sect. 12.3.5.2, the number of changes suggested by the ideality tool and the life principles tool were 2.05 and 1.22 respectively (Borderline significant, $p < 0.01$). There is no significant difference in the difficulty to suggest a change, as reported by the users. Hypothesis 3 is accepted partially. The sub-measure of 'Number of changes' indicates higher scores for the ideality tool (border line significance). The sub-measure of 'difficulty to suggest a change' did not indicate higher scores for the ideality tool.

Hypothesis 4 Students who are exposed to the ideality tool will get higher scores in the sub-measures of **Perceived usability** compared with students who are exposed to the life-principles tool.

We performed two sample t-test procedure (Table 12.18) to assess the differences between the ideality and life principle tools in the "Perceived usability" sub-measurers. No significant difference between the tools was found. Hypothesis 4 is rejected.

Discussion

We assessed the Ideality and life principles tools as sustainability tools for identification of sustainability principles and for sustainable design. First, we demonstrated that using these tools yielded higher scores in several measurers compared to a state of using no tool (Hypothesis 1). Students reported on higher levels of confidence, sustainability understanding and satisfaction from the sustainability analysis, as well as less difficulties in the analysis process, after using the tools. The

Table 12.18 T-test procedure to assess Hypothesis 4: perceived usability

		N	Mean	Std	Min	Max	Median	Two sample t-test
Difference between: It was easy to identify the sustainability principles of the biological system (After-Before)	Ideality	16	1.79	4.17	−7.9	6.7	3.35	ns*
	Life principles	17	2.86	3.96	−9.1	9.3	3.8	
Difference between: I felt confidence during the analysis (After-Before)	Ideality	16	2.77	3.15	−3.3	7.1	3.3	ns*
	Life principles	17	2.05	3.77	−9.2	7.3	2.7	
Difference between: I completely understood the sustainability principles of the biological system (After-Before)	Ideality	16	2.46	2.76	−2.8	6.1	2.25	ns*
	Life principles	17	1.78	3.6	−9.6	5.5	2.6	
Difference between: I am pleased from my sustainability analysis results (After-Before)	Ideality	16	1.83	2.85	−3.1	7.1	1.75	ns*
	Life principles	17	1.55	3.66	−9.7	7	1.7	
I will consider using the tool in the future in case of sustainability analysis	Ideality	19	6.07	2.6	0	9.2	6.9	ns*
	Life principles	19	7.47	1.8	2.8	9.9	7.7	

objective measure of number of sustainability principles identified is also significantly higher and divergent after using the tools. This result demonstrates the huge impact sustainability tools have in general. Some students were even helpless, searching for guidance at stage 2, as they did not know how to address the sustainability analysis process.

In relation to comparing the ideality and life principles tools we demonstrated several differences. First, in terms of sustainability principles identifications (Hypothesis 2), the ideality tool has higher validity as student's accuracy in replicating expert's identifications is higher and their missed identifications are lower, compared to the life principle tool. The reliability of the ideality tool is higher mainly when we consider repetition on right identifications (repetition 2), as most of the life principles students were correlated with their wrong identifications, thus their repetition rate in wrong identifications is relatively high. Even though the difference in reliability between the tools is not statistically significant, the size of the effect is relatively high (77.9 vs. 64.9). The significance of this effect is derived from the sample size and we assume that in larger samples it will turn out statistically significant.

In terms of sustainability design (Hypothesis 3), students who used the ideality tool offered in average more changes than students who used the life principle tool. This difference increased when we used the fixed data, considering the advantage of life principles in the number of principles. Even though this difference is only border line significant, the size of the effect is relatively high (2.05 vs. 1.22) and we assume that in a larger sample size, this effect will be statistically significant. However, students who offered changes by the ideality tool did not report on fewer difficulties in this process, compared to students who used the life principle tool.

In terms of student perceived usability (Hypothesis 4) there was no significant difference between the tools. Students reported on same levels of confidence, difficulties, sustainability understanding and satisfaction from the sustainability analysis, as well as tendency to use the tool in the future.

Apparently, the major difference between the tools is in the objective measures such as the sustainability identification validity/reliability and number of suggested changes. This difference between the tools may be explained by the nature of each one of the tools, presented in Sect. 9.4.3. Some of the life principles are more general while the ideality principles are more operative and descriptive. Life principles are based on a holistic view while the ideality principles are based on a technical view that might be more inherent to engineers. As a result, ideality students could more clearly identify the ideality principles as well as using these principles to offer changes, compared to the life principle students.

In addition, we have to bear in mind that we included only half of the life principles in this experiment, the half that more resembles the ideality principles and is also more clear and applicable to design. There are other life principles relating to the strategies of 'Evolve to survive', 'Development with growth' and 'Use friendly chemistry' that may be even less clear and descriptive, and incorporating them in this experiment might have even increased the difference between the tools.

As both students in stage 4 compared their sustainability analysis to stage 2, a stage of no tools, both of them felt major improvement. As a result, both tools demonstrate higher rates of perceived usability, and no tool has an advantage in terms of the improvement size.

Part V
Epilogue

Chapter 13
Discussion and Summary

13.1 Major Achievement—The Structural Biomimetic Design Method

This research yielded a new biomimetic design method that is based on structure-function patterns and incorporates TRIZ tools. The method is supported by a structured algorithm and auxiliary tools that lead the designer from biology to an application or vice versa. Major tools include the patterns table (Table 10.4) the ideality tool (Table 10.10), the complete viable system abstraction engine and brake models (Tables 10.7 and 10.8), and the transfer platform of analogy comparison components (Tables 10.11 and 10.12).

The method has been demonstrated by several case studied and experimented by lab and field experiments. These experiments demonstrated the potential of this design method to foster sustainability and innovation. The experiments also illuminated which innovations aspects are mainly provided by this design method.

13.2 Evaluation of Research Results Compared to Research Objectives

The main research gap identified in Chap. 4 is bridging the gap between biology and technology. The main biomimetic design stages suggested to bridge that gap (presented in Sects. 2.2 and 2.3) are intuitive and clear. Now it is about time to provide designers with auxiliary tools to follow these known stages, mainly tools that assist the search and abstraction stages of the biomimetic design process. The structural biomimetic design method indeed may be classified as an abstraction method, however, it has also the potential to be elaborated as a searching method, as described in Sect. 3.1, by using the "FindStructure" database as a source of

Y.H. Cohen and Y. Reich, *Biomimetic Design Method for Innovation and Sustainability*, DOI 10.1007/978-3-319-33997-9_13

biological solutions. In addition, the role of structure-function patterns as keywords for general searching in biological databases should be evaluated.

The method is based on the identified research gaps described in Sect. 4.1: Structure-function patterns, TRIZ knowledge base about systems and physical effects and nature sustainability patterns (integrated as part of the method). Thus, the method realizes the unfulfilled potential of the TRIZ knowledge base and patterns approach to contribute to the formation of the biomimetic discipline, extending other efforts described in Chap. 3.

We acknowledge that the main disadvantage of this structural biomimetic design method is that it leans on structures while in some cases complicated processes are involved that are not necessarily related to structures. Nevertheless, there are enough cases of structured base solutions, derived from the first and oldest strand of biomimetic, the functional morphology, form and function (Sect. 1.3).

13.3 Innovative Aspects of This Research

The innovative aspects of this research are mainly related to:

(i) Interdisciplinary knowledge transfer as a base for innovation—Engineering models derived from the TRIZ knowledge base infused our understanding about biological systems. The new gained knowledge about biological systems is in turn transferred back to engineering to evoke biomimetic innovations.

(ii) Viewing biological structures as part of a complete system model and through a technical lens—We used the complete viable system model [144] combining the TRIZ Su-Field model and the law of system completeness, to analyze biological systems. Engineers and designers who are familiar with this technical way of thinking might find this approach helpful in abstracting the complexity that characterizes structure-function relations in nature.

(iii) High level of structure-function patterns (Table 10.4) derived from biomimetic applications—These patterns span the space of possible structure-function relations and serve as "Front-end" index to thorough studies, providing the clues to open gates for zones of extended knowledge. The Findstructure database is an innovative biomimetic database, a source of biological systems classified and analyzed by theses structure-function patterns.

(iv) Classifying nature structures to engines or brakes—Our analysis yielded structures that are classified in the context of their **role** in the complete system; engines or brakes. This classification illuminates sustainability aspects of nature structures. Nature structures use free and clean energy sources like engines or provide defense against threats like brakes. This classification of nature structures to engines and brakes can inspire new breakthrough innovative design concepts, like a future engine that is based on a structure that exploits, at least partially, nature forces for population instead of fossil fuels.

(v) Ideality tool for sustainability and critical thinking—The ideality tool provides an answer for the real need of sustainability tools during early concept design stages. In relation to biomimetic design, the ideality tool enriches the life principles knowledge base and provides an operative framework intuitive to engineers, addressing sustainability aspects as **integral** part of the biomimetic design process. Ideality is also a tool for critical thinking to observe nature imperfections (Sect. 1.7) and identify possible improvements to the ideality of the system.

13.4 By-Product Contributions (Added Value)

There are two main by-product contributions to the design literature and to the innovation literature.

(i) Design literature—This research documents a development of a new design method which may be valuable for the development of other design methods. The research model (Chap. 5, Fig. 5.3) reflects our design philosophy of iterative study and development [126] and guides the construction of knowledge required for a new design method. This documented study in general, and the research model in particular contributes to the empirical aspect of the design literature.

(ii) Innovation criteria—This research yielded innovation criteria rubric to assess innovation aspects of design results, adjusted for evaluation of design concepts (Table 12.2). The rubric was elaborated and adjusted for this study based on several rubrics from the literature. Thus, the new rubric enriches the innovation criteria literature and can be used for other innovation assessments purposes.

13.5 Future Research

The following have been identified as promising future research topics:

- Continuous identification of Ideality patterns by increasing the scope of analysis —So far we mainly observed biological structures; it may be interesting to observe also the ideality patterns of biological processes and ecosystems and assess the ideality tool in other contexts, both in design classes and in industry. The integration of the ideality tool within a biomimetic design process should be further discussed and demonstrated by case studies.
- Continuous identification of structure-function patterns—It may be interesting to search for new structure-function patterns by using different system modeling approaches such as the SAPPhIRE model [112] or functional modeling [85]. In addition, further research should focus on thorough studies revealing the detailed mechanisms behind these meta-level patterns.

- Using structure-function patterns as keywords for organism search—Most of the searching methods are based on functional keywords. Compared to functions, structures are less subjected to personal interpretations and to domain terminologies. While biological and technological functions are derived from different terminologies, structures are visual and therefore less subordinated to different interpretations. Adding structures to functional keywords may lead to more qualified and accurate results. However, our suggested generic structures are too abstract to lead the search per se, as they may appear in different ways in biological texts. The searching terms would be more beneficial if we add details to the structure description such as the type of protrusion, i.e., hairs, denticles, bristles etc. Further studies could elaborate on relevant structural words in biological texts. Thus we may elaborate the structural biomimetic design method to be also a searching method in general databases.
- "FindStructure" Database—Further development of the Findstructure database by adding more records of biological functional mechanisms analyzed by the structure-function patterns. Thus, we elaborate the structural biomimetic design method to be also a searching method. Further assessment of the database as a supporting tool for biomimetic design processes in general and the biomimetic design method in particular, is recommended.
- Analyze the structural biomimetic design method with a design theory such as C-K theory [107, 225] or a framework such as PSI [67–69], in the same way that was done with ASIT [226] or Infused design [124] and obtain insight about the method and the theory, with potential new opportunities to extend the method.
- Evaluate the structural biomimetic design method by industrial case studies.
- Evaluate the structural biomimetic design method worldwide among various target users including biologists, engineers, designers and more.

Appendix A
Innovation Experiment—Biological System 1

Some geckos have the ability to voluntary shed their tail as a defensive mechanism against predators. The lizard can shed its tail when it is grasped by a predator or when it feels threatened. The detached tail flutters, attracts the predator's attention and let the lizard escape safely. A new tail is growing up within weeks to months.

The shedding of the tail always occurs in a distinct area of a fracture plane. This area is constructed of cartilage and contains minor blood vessels to minimize the blood lost during shedding. The fracture plane is identified with "score lines" that creates a weakness area. During the detachment of the tail, the fracture plane opens like a zipper, creating a "crown" of zipper protrusions as demonstrated in Fig. A.1a. Fig. A.1b is an enlargement of one zipper protrusion. This shedding mechanism needs to be rapid and therefore it is not based on chemical process of proteolysis, but on a mechanical structural process.

The two parts of the tail, before and after the fracture plane, are attached by biological adhering forces. Before the shedding, microstructures at the terminal end of the muscle fibers, has a shape of rods that enlarge the surface area and facilitate the biological adhesion. During the shedding, the lizard contracts the muscles around the fracture plane, and as a result, the shape of the rods is changed into "mushroom-shaped" structures (Fig. A.1c). Consequently, the interaction area of the biological adhesion forces at the fracture plane is lowered, facilitating autotomy.

Fig. A.1 End of lizard tail after the shedding. Adapted from [208] under CC Attribution License. **a** End of lizard tail after the shedding—the "zipper" crown structure. **b** Zoom on one zipper protrusion. **c** Scanning electron microscopy of one "zipper" protrusion: Muscles terminals have a "mushroom-shaped" structure

Y.H. Cohen and Y. Reich, *Biomimetic Design Method for Innovation and Sustainability*, DOI 10.1007/978-3-319-33997-9

The muscle contractions are also likely to facilitate the removal of the skin and muscles to complete the release of the tail.

This shedding process is initiated by the predator that grasp the tail or by the lizard itself when it feels danger. In the last case, the nerve system initiates the process by contracting the muscles at the fracture plane.

Appendix B
Innovation Experiment—Biological System 2

Many seeds resist water penetration. The Papilionaceae seed is kept dry by a humid reduction mechanism that keeps the humid levels inside the seed lower or equal to the humid levels in the environment.

When the seed is detached from the plant, a scar is remained in the detachment area. This scar is actually a hilum that functions as a unidirectional valve. Humid can be released in one direction, from the seed outside, supporting the desiccation process of the seed.

In the long axis of the hilum there are two epidermis layers. Each one is constructed of pillar cells vertically to the epidermis layer (Fig. B.1a). An impermeable tissue separates the external and internal epidermis layers. As a result, the external epidermis layer can react only to the humid levels in the environment whereas the internal layer reacts to the humid levels within the seed.

The pillar cells that construct the epidermis layers react to humidity by absorbing or desiccating humid. When a pillar cell absorbs humid it is swelled and when it releases humid it is shrunk. The change in volume in these pillar cells creates a tension and loose movement as a result of the exerted hygroscopic pressure.

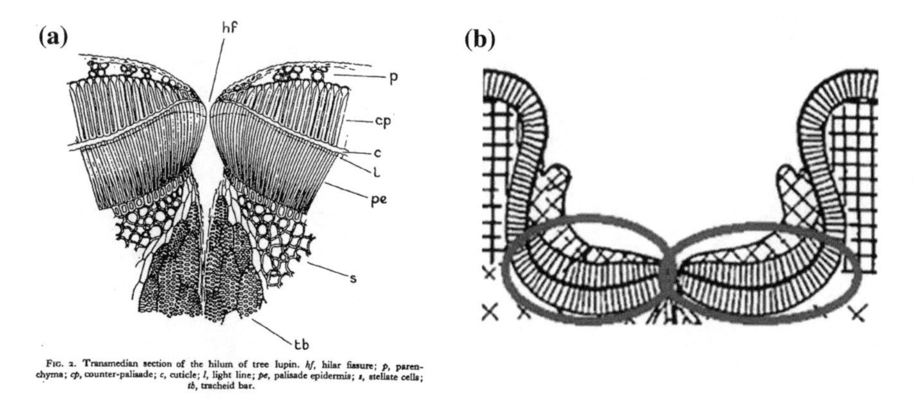

Fig. B.1 The Papilionaceae seed: schematic figure of the scar and the hilum area. **a** schematic description of the hilum area itself. **b** schematic figure of the scar location on the seed envelope. Image adapted from [207], reproduced with permission of the copyright owner

Y.H. Cohen and Y. Reich, *Biomimetic Design Method for Innovation and Sustainability*, DOI 10.1007/978-3-319-33997-9

In addition, the pillar cells that define the height of the epidermis are distributed asymmetrically as the pillars are getting shorten when we approach the hilum (Fig. B.1b), having a concave shape. As a result, when the external epidermis cells absorb more humid comparing to the cells in the internal epidermis layer (when the humid level in the environment is higher than the humid level within the seed), they swell and extract pressure that close the hilum. When the internal epidermis cells absorb more humid comparing to the cells in the external epidermis layer (when the humid level within the seed is higher than the humid level outside the seed), they swell and extract pressure that opens the hilum and release humid.

Appendix C
Innovation Experiment—Student Form

What is the main challenge solved by the described biological system (Gecko / Papilionaceae seed)?

____

Identify an analogical design challenge:

____

____

Suggest a design concept to solve your identified challenge, based on the Gecko / Papilionaceae seed system:

____

____

____

____

Describe <u>in details</u> the operative mechanism of the system you suggest to design in a way that your team colleagues will be able to understand it without your help.

____

____

____

____

Describe your solution by a sketch. Include all the details mentioned literally before:

Y.H. Cohen and Y. Reich, *Biomimetic Design Method for Innovation and Sustainability*, DOI 10.1007/978-3-319-33997-9

Appendix D
Structural Modeling Template Form

Analyzing the Gecko/Papilionaceae seed system by the structural biomimetic abstraction method:

1. Identify structures from the patterns table that appear at the system description. There might be more than one structure. What is their related function?

____

____

	Structural pattern	**Types / private cases**	**Generic function**	**Generic functions (second hierarchy)**	**Generic functions (third hierarchy)**
1	Repeated protrusions		Move **(Engines)**	Attach Detach	Connect, Combine, Join, Adhere, Bond, Add, Increase. Remove, Subtract, Decrease.
2	Repeated tubes / channels / tunnels	Without valves		Channel	Lead, Guide, Direct, Flow, Stream, Transfer.
		With Valves		Regulate	Control, Modulate, Separate, Filter.
3	Asymmetry	Geometric Asymmetry		Change	Change position or location: Rotate, Spin, Turn, Move up, Move down, Move aside, Open, Close.
		Material Asymmetry			
		Time Asymmetry			Change volume or form: Blow, Blast, Cut.
4	Layers (Sandwich)		Stop **(Brakes)**	Protect or defend against mechanical or thermal loads.	Absorb, Push back, Resist, Isolate, Insulate (heat).
5	Intersected layers	Network, cellular, honeycomb			
6	Tube				
7	Helix				
8	Streamlined shapes	Spiral, beak & body contours		Protect or defend against dynamic loads (turbulences).	Stabilize, Disperse, Deflect, Smoothen.
9	Container	Sphere, cups		Protect or defend against gravitation / mechanical loads.	Contain, Store, Hold, Grasp, Trap.

Y.H. Cohen and Y. Reich, *Biomimetic Design Method for Innovation and Sustainability*, DOI 10.1007/978-3-319-33997-9

2. Construct a system model for each one of the structures you identified, according to the type of structure: Engine/Brake. You can use the following templates:

Fill in for the gecko's tail / Papilionaceae seed system (Engine model):

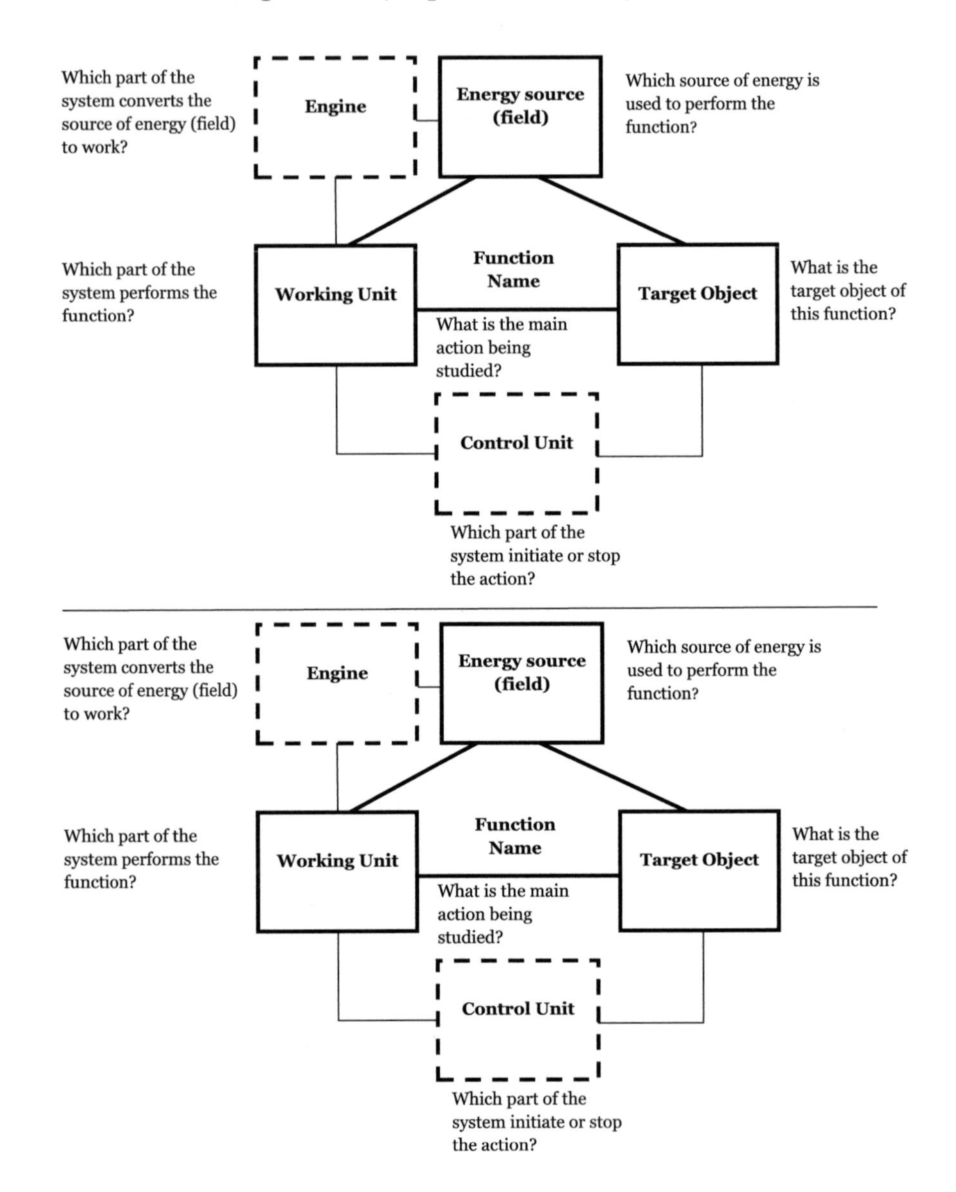

Fill in for the gecko's tail system / Papilionaceae seed system (Brake model):

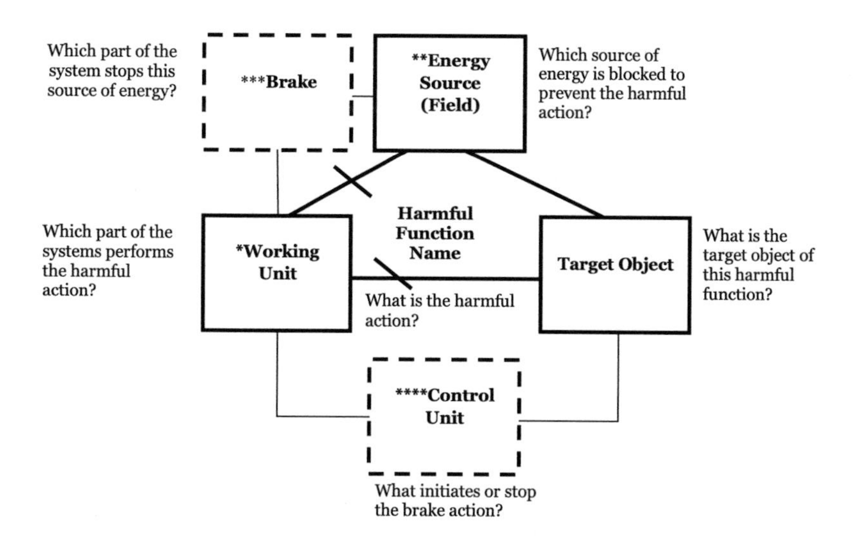

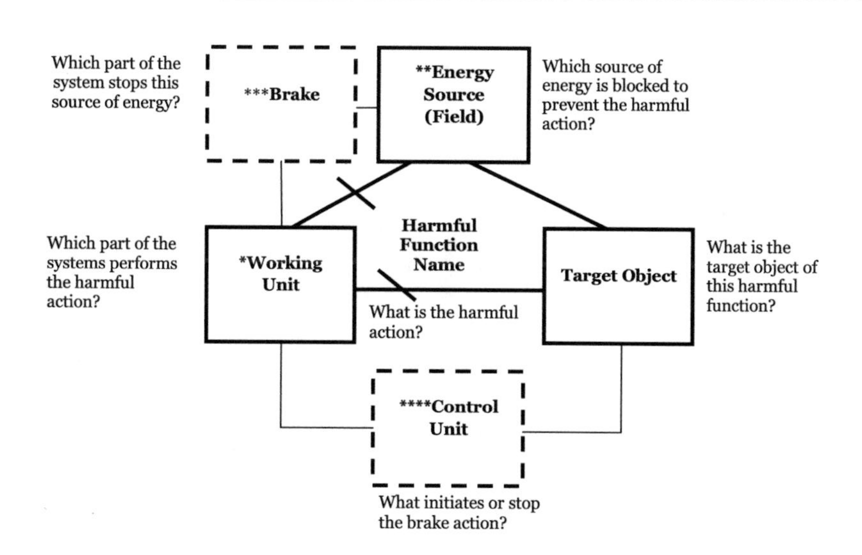

Appendix E
Innovation Assessment Criteria Form

	Measure	Statement	1	2	3	4	5	6	7	8	9	10
1	Novelty	Uniqueness - The proposed design concept is original (unusual).										
		Surprise - The proposed design concept is surprising (unexpected).										
2	Usefulness	Value - The proposed design concept solves a valuable problem.										
		Functionality - The proposed design concept achieves its declared function, as defined by the subjects (solve the problem).										
		Feasibility - The proposed design concept is feasible in a reasonable cost-effective ratio.										
		Patentability - The proposed design concept may be patentable.										
3	Cohesiveness	Cohesiveness - The proposed design concept appears complete, well developed and is detailed enough to be understandable.										
4	General Innovation	The proposed design concept is innovative according to my experience as innovation consultant.										
		The proposed design concept is a breakthrough innovation offering a paradigm shift.										
5	Relative Innovation	The proposed design concept is innovative relative to other solutions in market.										

Y.H. Cohen and Y. Reich, *Biomimetic Design Method for Innovation and Sustainability*, DOI 10.1007/978-3-319-33997-9

Appendix F
Sustainability Analysis—Biological System 1

The desert snail [87, 222] can survive the dessert extreme climate, where the surface temperature can reach 70 °C and more.

The snail is protected by a shell, a stiff and essential structure for the snail. The shell protects against predators and provides a dried refuge. The shell is made of calcium carbonate and constructed of several spirals.

The snail survives the extreme high temperatures due to the shell heat transfer mechanism that is based on two basic principles (Fig. F.1):

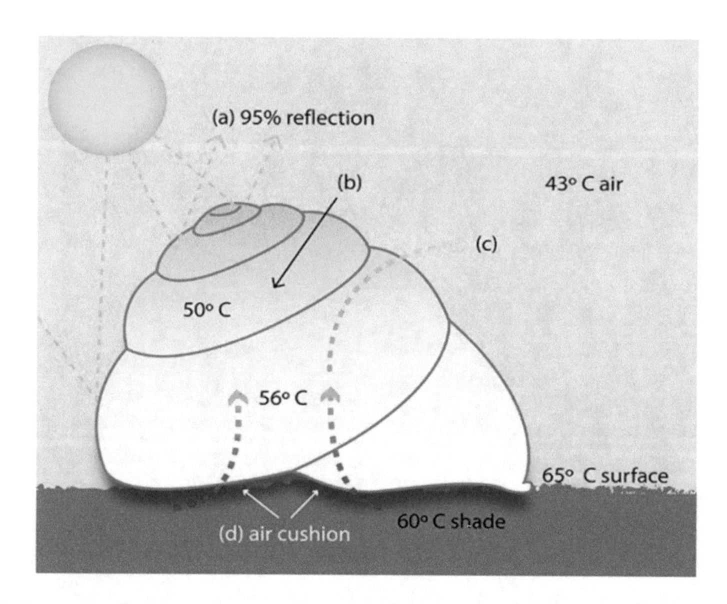

Fig. F.1 Image by Gretchen Hooker, Biomimicry 3.8 Institute. Reproduced with permission

Y.H. Cohen and Y. Reich, *Biomimetic Design Method for Innovation and Sustainability*, DOI 10.1007/978-3-319-33997-9

(1) High reflectivity of solar radiation—The shell reflects well over 95 % of the solar radiation. As a result of this high reflectivity, the radiation heat flow does not reach the snail.
(2) Slow conduction of heat from the substrate:

- The snail has only few spots of direct contact with ground, forming an insulating air cushion.
- The snail is withdrawn into the upper spiral of the shell and the lower largest spiral is filled with air that serves as another impediment to heat flow.

As a result of temperature differences, heat flows by conduction from the ground to the surrounding air through the first spiral of the shell, and not through the second spiral where the snail dwells.

Appendix G
Sustainability Analysis—Biological System 2

The water fern Salvinia Molesta [223, 224] (Fig. G.1a) can be covered with an air layer under the water. If the fern is pressed under water by wind or animals, it quickly builds up an air layer that prevents a contact between the fern and the water and let it stayed dried during weeks in water.

The air layer is stayed stable also in turbulent flow environment. The fern continues respiration and do not submerge in water due to the floating air layer. The air layer reduces the shear stress and the surface friction drag of the fern surface. When the fern goes up from the water it is stayed dried with no wetting signs on its surface.

The fern leaves are covered with dense elastic hairs (Fig. G.1b). Every four hairs are grouped together in an eggbeater shape complex (Fig. G.1c) and project 2 mm from the leave surface. This eggbeater structure that identifies the distanced edge of the hairs is hydrophilic (attracts water). The leave surface itself is covered with waxy hydrophobic layer (repel water). Water droplets are detached from the

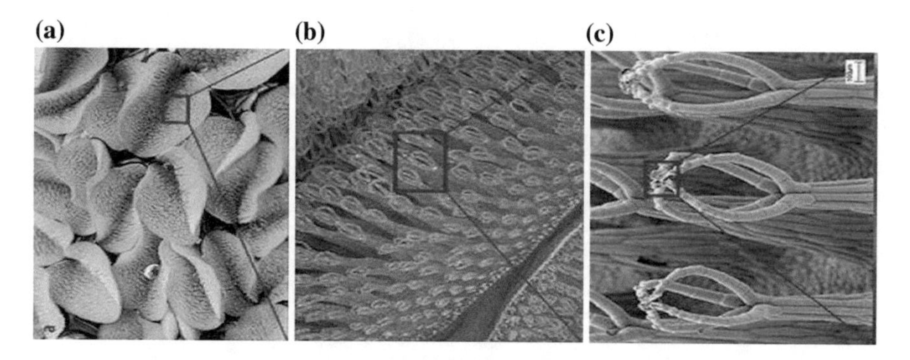

Fig. G.1 **a–c** From Wikipedia, based on Barthlott et al. 2010 [223]

Y.H. Cohen and Y. Reich, *Biomimetic Design Method for Innovation and Sustainability*, DOI 10.1007/978-3-319-33997-9

hydrophobic waxy layer and get trapped and fixed at the hydrophilic eggbeater hairs edges. The required energy to move this trapped water droplet by air bubbles is very high, so the air bubbles are actually trapped between the hydrophobic waxy leave surface and the hydrophilic edge of hairs. The result is a long-term air-retention.

This air retention mechanism is called the Salvinia Effect.

Appendix H
Students Instructions Before Exposure to Sustainability Tools

What makes this system (the desert snail/Salvinia) a sustainable system? What are the sustainability principles of this system?

____

____

____

____

____

Y.H. Cohen and Y. Reich, *Biomimetic Design Method for Innovation and Sustainability*, DOI 10.1007/978-3-319-33997-9

Appendix I
Students Instructions After Exposure to Ideality Tool

What makes this system a sustainable/ideal system? Use the following table. Give an example for every ideality design principle that is demonstrated by the dessert snail/Salvinia system.

Y.H. Cohen and Y. Reich, *Biomimetic Design Method for Innovation and Sustainability*, DOI 10.1007/978-3-319-33997-9

Increasing benefits:

General Strategy	Design Principle	Example – Desert Snail / Salvinia
More Functions	Multifunctional design – enlarge the number of functions related to one structure by unification of system parts.	
Stronger Effect – Intensify function values	Intensify the interaction with the environment to achieve stronger effect of one function by: ✓ Repetition on elements ✓ Increased surface area	

Reduction of costs:

General Strategy	Design Principle	Example – Desert Snail/ Salvinia
Defensive Strategy *Prevent disturbances and damages to the system*	Reduction of disturbances such as friction, loads, turbulence and more by structures.	
	Decrease of surface area when it is harmful effect.	
Opportunistic Strategy *Usage of available sources to save cost*	Usage of physical, chemical, geometrical and other effects and gradients as energy sources.	
	Adjustment of structure to function – Structure provides the function.	
	Transfer some functions to the supersystem. Using supersystem material resources.	
Prevent Waste *For better usage of resources*	Synchronize system parameters to prevent waste.	
	Improve the conductivity of energy through the system to provide easier access and prevent waste of energy.	
	Give-up redundant parts.	

Assume that you can perform changes in the desert snail/Salvinia system. Offer as many design changes as possible to the desert snail/Salvinia system to increase its sustainability (That more ideality principles will exist at the dessert snail/Salvinia system).

___

___

___

___

Appendix J
Students Instructions After Exposure to Life Principles Tool

What makes this system a sustainable system? Use the following table. Give an example for every life principle that is demonstrated by the dessert snail/Salvinia system.

General Strategy	Design Principle	Example– Desert Snail / Salvinia
Be Resourceful – Material & Energy Efficient *Organisms survives when they use resources efficiently*	Use Multifunctional Design.	
	Using available energy sources.	
	Recycle all materials.	
	Fit form to function.	
Adapt to changing conditions *Organisms survive when they are adjusted to changing environment*	Self – Renewal.	
	Incorporate Diversity.	
	Redundancy- The same part in system is duplicated many times.	
	Decentralization – different parts at the system have the same function.	
Be Locally attuned and responsive *Organisms "Listen", attached and respond to their environment*	Use readily available Local sources.	
	Use feedback loops.	
	Leverage cyclic processes.	
	Cultivate cooperative relationship.	

Y.H. Cohen and Y. Reich, *Biomimetic Design Method for Innovation and Sustainability*, DOI 10.1007/978-3-319-33997-9

Assume that you can perform changes in the desert snail/Salvinia system. Offer as many design changes as possible to the desert snail/Salvinia system to increase its sustainability (That more life principles will exist at the dessert snail/Salvinia system).

____

____

____

____

Appendix K
Expert's Analysis by Ideality Tool—Desert Snail

Enlarging benefits:

General Strategy	Design Principle	Desert Snail
More Functions	Multifunctional design – enlarge the number of functions related to one structure by unification of system parts.	Shell has multiple functions: ✓ Shelter for snail against predators ✓ Dried area ✓ Temperature regulation **1**
Stronger Effect – Intensify function values	Intensify the interaction with the environment to achieve stronger effect of one function by: ✓ Repetition on elements ✓ Increased surface area	✓ Repetition of elements: Several spirals – larger air layers -better heat transfer. ✓ Repetition of heat regulation mechanisms- (i) High reflectivity (ii) Slow conduction of heat. **1**

Reduction of costs:

General Strategy	Design Principle	Desert Snail
Defensive Strategy *Prevent disturbances and damages to the system*	Reduction of disturbances such as friction, loads, turbulence and more by structures.	✓ Reduction of heat by behavior: climbing up to create an air layer ✓ Stiff shell sustain mechanical damage **1**
	Decrease of surface area when it is harmful effect.	The snail has only few spots of direct contact with ground, forming an insulating air cushion. **1**
Opportunistic Strategy *Usage of available sources to save cost*	Usage of physical, chemical, geometrical and other effects and gradients as energy sources.	Using temperature gradient to transfer heat **1**
	Adjustment of structure to function – Structure provides the function.	✓ Spiral adjusted for going up, for heat transfer. ✓ Shell is adjusted for reflecting radiation. **1**
	Transfer some functions to the supersystem. Using supersystem material resources.	Using air for heat transfer. **1**
Prevent Waste *For better usage of resources*	Synchronize system parameters to prevent waste.	Going up is synchronized with external temperature to prevent waste of energy for cooling **1**
	Improve the conductivity of energy through the system to provide easier access and prevent waste of energy.	✓ The spiral shape directs the heat flow away from the snail. ✓ Shell reflects radiation. **1**
	Give-up redundant parts.	– **0**

Y.H. Cohen and Y. Reich, *Biomimetic Design Method for Innovation and Sustainability*, DOI 10.1007/978-3-319-33997-9

Appendix L
Expert's Analysis by Ideality Tool—Salvinia Fern

Increasing benefits:

General Strategy	Design Principle	Salvinia
More Functions	Multifunctional design – enlarge the number of functions related to one structure by unification of system parts.	Air layer has several functions: ✓ Respiration. ✓ Floating – prevent submerging. ✓ Reduce friction. ✓ Reduce shear stress. **1**
Stronger Effect – Intensify function values	Intensify the interaction with the environment to achieve stronger effect of one function by: ✓ Repetition on elements ✓ Increased surface area	✓ Repetition of hairs- stronger effect of water removal. **1**

Reduction of costs:

General Strategy	Design Principle	Salvinia
Defensive Strategy *Prevent disturbances and damages to the system*	Reduction of disturbances such as friction, loads, turbulence and more by structures.	✓ Reduction of water disturbance - (decomposition) ✓ Reduction of friction, shear stress **1**
	Decrease of surface area when it is harmful effect.	– **0**
Opportunistic Strategy *Usage of available sources to save cost*	Usage of physical, chemical, geometrical and other effects and gradients as energy sources.	✓ Using chemical gradient (hydrophilic- hydrophobic) to move water. ✓ Can use water movement to spread away as it floats. **1**
	Adjustment of structure to function – Structure provides the function.	Hairs and hairs tips are adjusted to the water removal function. **1**
	Transfer some functions to the supersystem. Using supersystem material resources.	Using air to remove water. **1**
Prevent Waste *For better usage of resources*	Synchronize system parameters to prevent waste.	– **0**
	Improve the conductivity of energy through the system to provide easier access and prevent waste of energy.	– **0**
	Give-up redundant parts.	– **0**

Y.H. Cohen and Y. Reich, *Biomimetic Design Method for Innovation and Sustainability*, DOI 10.1007/978-3-319-33997-9

Appendix M
Expert's Analysis by Life Principles Tool—Desert Snail

General Strategy	Design Principle	Desert Snail
Be Resourceful – Material & Energy Efficient *Organisms survives when they use resources efficiently*	Use Multifunctional Design.	Shell has multiple functions: ✓ Shelter for snail against predators ✓ Dried area. ✓ Temperature regulation **1**
	Using available energy sources.	Using temperature gradient to transfer heat. **1**
	Recycle all materials.	Shell is decomposed. **1**
	Fit form to function.	✓ Spiral is fitted for going up and for heat transfer. ✓ Shell is fitted to reflect radiation. **1**
Adapt to changing conditions *Organisms survive when they are adjusted to changing environment*	Self – Renewal.	Applying behavioral "repair" to changing temperatures by going up. **1**
	Incorporate Diversity.	– **0**
	Redundancy- The same part in system is duplicated many times.	✓ Several spirals – larger air layers -better heat transfer. **1**
	Decentralization – different parts at the system have the same function.	Two system parts provide the function of heat regulation: (i) High reflectivity (ii) Slow conduction of heat. **1**
Be Locally attuned and responsive *Organisms "Listen", attached and respond to their environment*	Use readily available Local sources.	✓ Using air for heat transfer. ✓ Shell is manufactured from calcium, local resource. **1**
	Use feedback loops.	✓ Snail goes up when it is hot, reacts to the external temperature. **1**
	Leverage cyclic processes.	– **0**
	Cultivate cooperative relationship.	– **0**

Y.H. Cohen and Y. Reich, *Biomimetic Design Method for Innovation and Sustainability*, DOI 10.1007/978-3-319-33997-9

Appendix N
Expert's Analysis by Life Principles Tool—Salvinia Fern

General Strategy	Design Principle	Salvinia
Be Resourceful – Material & Energy Efficient *Organisms survives when they use resources efficiently*	Use Multifunctional Design.	Air layer has several functions: ✓ Respiration. ✓ Floating – prevent submerging. ✓ Reduce friction. ✓ Reduce shear stress. **1**
	Using available energy sources.	✓ Using chemical gradient (hydrophilic-hydrophobic) to move water. ✓ Can use water movement to spread away as it floats. **1**
	Recycle all materials.	✓ Salvinia is decomposed. ✓ Water and air are returned to the ecosystem when the fern goes up from the water. **1**
	Fit form to function.	Hairs and hairs tips fit to the removal function **1**
Adapt to changing conditions *Organisms survive when they are adjusted to changing environment*	Self – Renewal.	✓ The air layer is stayed stable also in turbulent flow environment, build and repair the layer as needed. ✓ New leaves can grow **1**
	Incorporate Diversity.	– **0**
	Redundancy- The same part in system is duplicated many times.	Many hairs. **1**
	Decentralization – different parts at the system have the same function.	Two parts have the function of water removal: 1. Hydrophobic layer 2. Hydrophilic tip **1**
Be Locally attuned and responsive *Organisms "Listen", attached and respond to their environment*	Use readily available local sources.	Using air to remove water. **1**
	Use feedback loops.	Feedback to water state - go over again and again. **1**
	Leverage cyclic processes.	– **0**
	Cultivate cooperative relationship.	– **0**

Y.H. Cohen and Y. Reich, *Biomimetic Design Method for Innovation and Sustainability*, DOI 10.1007/978-3-319-33997-9

Appendix O
Sustainability Analysis Students Questionnaire (After Stage 2)

Age: ______

Gender: M / F

	Few	Extensive
Previous experience in sustainable design		
Pervious knowledge in sustainability		

		Disagree	Agree
1	It was easy to identify the sustainability design principles of the biological system.		
2	I felt confidence during the sustainability analysis process.		
3	I completely understood the sustainability principles of the biological system.		
4	I my satisfied with my sustainability analysis results.		

Y.H. Cohen and Y. Reich, *Biomimetic Design Method for Innovation and Sustainability*, DOI 10.1007/978-3-319-33997-9

Appendix P
Sustainability Analysis Students Questionnaire (After Stage 4)

		Disagree — Agree
1	It was easy to identify the sustainability design principles of the biological system.	
2	I felt confidence during the sustainability analysis process.	
3	I completely understood the sustainability principles of the biological system.	
4	I felt difficulty to offer design changes to the biological system to enlarge its sustainability (more ideality principles / life principles will exist).	
5	I my satisfied with my sustainability analysis results.	
6	I will consider using Ideality / life principle tools in the future (if I need to address sustainability issues).	

Y.H. Cohen and Y. Reich, *Biomimetic Design Method for Innovation and Sustainability*, DOI 10.1007/978-3-319-33997-9

References

1. Benyus, J. *M., Biomimicry: Innovation Inspired by Nature*. Quill. New York, 1997.
2. Shu, L., Ueda, K., Chiu, I., Cheong, H., Biologically inspired design. *CIRP Annals-Manufacturing Technology*, **60**(2): p. 673–693, 2011.
3. Nachtigall, W., *Bionik: Grundlagen und Beispiele für Ingenieure und Naturwissenschaftler*. Springer DE, 2002.
4. Gleich, A., Pade, C., Petschow, U., Pissarskoi, E., *Potentials and Trends in Biomimetics*. Springer, 2009.
5. http://biomimicry.net/. [cited 01.25.2014].
6. Steele, J., How do we get there? Bionics Symposium: Living Prototypes The Key to New Technology, in WADD Technical Report No. 60-600 Wright Air Development Division, Wright-Patterson Air Force Base. 1960.
7. Merrill C.L., *Biomimicry of the Dioxygen Active Site in the Copper Proreuns Hemocyanin and Cytochrome Oxidase*, PHD Thesis, Rice University, 1982.
8. Bejan, A., *Shape and Structure, From Engineering to Nature*. Cambridge University Press, 2000.
9. http://www.pointloma.edu/experience/academics/centers-institutes/fermanian-business-economic-institute/da-vinci-index-biomimicry. [cited 12.23.2013].
10. Fermanian Business & Economic Institute, P.L.N.U., *Bioinspiration: an economic progress report*, 2013.
11. Lurie-Luke, E., Product and technology innovation: What can biomimicry inspire? *Biotechnology advances*, 32(8): p. 1494–1505, 2014.
12. http://www.naturaledgeproject.net/Keynote.aspx. [cited 12.23.2013].
13. Ullman, D.G., *The Mechanical Design Process*. 4th ed. New York: McGraw–Hill, 2009.
14. Reich, Y., Preventing Breakthroughs from Breakdowns, *In Proceedings of the 9th Biennial ASME Conference on Engineering Systems Design and Analysis (ESDA)*, Haifa, Israel, 2008.
15. Vosniadou, S., Analogical reasoning as a mechanism in knowledge acquisition: A developmental perspective. *Similarity and analogical reasoning*, p. 413–437, 1989.
16. Goel, A.K., Design, analogy, and creativity,*IEEE expert*, **12**(3), p. 62–70, 1997.
17. Mak, T., and Shu, L., Abstraction of biological analogies for design. *CIRP Annals-Manufacturing Technology*, **53**(1), p. 117–120, 2004.
18. Schild, K., Herstatt, C., and Lüthje, C., *How to use analogies for breakthrough innovations*. Working Papers/Technologie-und Innovationsmanagement, Technische Universität Hamburg-Harburg, 2004.
19. Benami, O., and Y. Jin., Creative stimulation in conceptual design. *In Proc. of ASME Design Engineering Technical Conference and* Computers and Information in Engineering Conference (DETC/CIE), Montreal, QC, Canada, **2**(1), 2002.

Y.H. Cohen and Y. Reich, *Biomimetic Design Method for Innovation and Sustainability*, DOI 10.1007/978-3-319-33997-9

20. Yen, J., Helms, M., Goel, A., Tovey, C., Weissburg, M., Adaptive Evolution of Teaching Practices in Biologically Inspired Design, in *Biologically Inspired Design*. Springer. p. 153–199, 2014.
21. Johansson, F., *The Medici effect: Breakthrough insights at the intersection of ideas, concepts, and cultures*. Harvard Business Press, 2004.
22. Turner, J.S., and Soar, R.C., Beyond biomimicry: What termites can tell us about realizing the living building. in *Proc. 1st Int. Conf. Industrialized, Intelligent Construction*, Loughborough University, UK, 2008.
23. Altshuller, G., *The Innovation Algorithm, TRIZ, Systematic Innovation and Technical Creativity*. Worcester, MA: Technical Innovation Center, Inc., 1999.
24. Vincent, J.F., Bogatyreva, O. A., Bogatyrev, N. R., Bowyer, A., Pahl, A. K., Biomimetics: its practice and theory. J R Soc Interface, 3(9), p. 471–82, 2006.
25. Fratzl, P., Biomimetic materials research: what can we really learn from nature's structural materials?. *Journal of the Royal Society Interface*, **4**(15), p. 637–642, 2007.
26. Suh, N.P., The Principles of Design. Oxford University Press. New York, 1990.
27. Brundtland, G.H., *Our Common Future: The World Commission on Environment and Development*, Oxford University Press, Oxford, 1987.
28. Hawken, P., Lovins, A.B., and Lovins, L.H., *Natural Capitalism: the Next Industrial Revolution*. Earthscan, 2010.
29. http://www.naturalstep.org/. [cited 01.15.2014].
30. Nidumolu, R., C.K. Prahalad, and M. Rangaswami, Why sustainability is now the key driver of innovation. *Harvard business review*, 87(9): p. 56–64. 2009.
31. Anderson, R.C., *Confessions of a radical industrialist*. New York/London: Random House Business Books, 2009.
32. Tukker, A., Life cycle assessment as a tool in environmental impact assessment. *Environmental Impact Assessment Review*, **20**(4), p. 435–456, 2000.
33. http://www.naturalstep.org/sites/all/files/TNS-SLCA-tool.pdf. [cited 17.02.14].
34. www.prosa.org. [cited 17.02.14].
35. http://www.informationinspiration.org.uk. [cited 17.02.14].
36. Fitzgerald, D.P., Herrmann, J.W ., Sandborn, P. A., Schmidt, L.C., Beyond tools: a design for environment process. *International Journal of Performability Engineering*, **1**(2), p. 105, 2005.
37. Charter, M., and Tischner, U., *Sustainable solutions: developing products and services for the future*. Greenleaf publishing, 2001.
38. Reap, J., Baumeister, D., and Bras, B., Holism, biomimicry and sustainable engineering. In *Proc. of International Mechanical Engineering, Congress and Exposition* (IMECE), Orlando, USA, 2005.
39. Antony, F., Grießhammer, R., Speck, T., Speck, O., Sustainability assessment of a lightweight biomimetic ceiling structure. *Bioinspiration & Biomimetics*, **9**(1), p. 016013, 2014.
40. Faludi, J., Biomimicry place in green design, *Zygote Quarterly*. p. 120–129, 2012.
41. Hoeller, N., Salustri, F., DeLuca, D., Pedersen, Z., Love, M., McKeag, T., Stephers, F., Reap, J., Sopchac, L., Patterns from Nature, in *Proc. of the SEM Annual Conference and Exposition on Experimental and Applied Mechanics*, Massachusetts, USA, 2007.
42. Reap, J., *Holistic Biomimicry: a Biologically Inspired Approach to Environmentally Benign Engineering*. PHD Thesis, Georgia Institute of Technology, 2009.
43. Zari, M.P., *Biomimetic approaches to architectural design for increased sustainability*. School of architecture, Victoria University, NZ, 2007.
44. Goel, A.K., Bras, B., Helms, M., Rugaber, S., Tovey, C., Vattam, S., Weissburg, M., Wiltgen, B., Yen, J., Design Patterns and Cross-Domain Analogies in Biologically Inspired Sustainable Design. *AAAI Spring Symposium Series*, 2011.
45. McDonough, W., and Braungart, M., *Cradle to cradle. Remaking the way we make things*. North Point Press, 2002.

46. Tempelman, E., Van der Grinten, B., Mul, E.J., De Pauw, I. Nature Inspired Design: a practical guide towards positive impact *products*. Enschede, Boekengilde, 2015.
47. Miller, K.R., Life's grand design. *MIT's Technology Review,* 97(**2**), p. 25–32, 1994.
48. Shai, O., Reich, Y., and Rubin, D., Creative conceptual design: extending the scope by infused design. *Computer-Aided Design*, **41**(3), p. 117–135, 2009.
49. *Engineering Biology for the 21st Century. A Plan for Bioengineering at Harvard*. Harvard University Bioengineering, 2008.
50. Bartlett, M.D., Croll, A.B., King, D.R., Paret, B.M., Irschick, D.J., Crosby, A.J., Looking beyond fibrillar features to scale gecko like adhesion. *Advanced Materials*, **24**(8), p. 1078–1083, 2012.
51. Bhushan, B., J.N. Israelachvili, and U. Landman, Nanotribology: friction, wear and lubrication at the atomic scale. *Nature*, 374(6523): p. 607–616, 1995.
52. Konda, S., Monarch, I., Sargent, P., Subrahmanian, E., Shared memory in design: a unifying theme for research and practice. *Research in Engineering design*, 4(1): p. 23–42, 1992.
53. Solga, A., Cerman, Z., Striffler, B. F., Spaeth, M., Barthlott, W., The dream of staying clean: Lotus and biomimetic surfaces. *Bioinspiration & Biomimetics,* **2**(4), p. S126, 2007.
54. Helms, M., Vattam, S.S., and Goel, A.K., Biologically inspired design: process and products. *Design Studies*, **30**(5), p. 606–622, 2009.
55. Hesselberg, T., Biomimetics and the case of the remarkable ragworms. *Naturwissenschaften*, **94**(8), p. 613–621. 2007.
56. Speck, T., Speck, O., Beheshti, N., McIntosh, AC., Process sequences in biomimetic research. *Design and Nature*, 4, p. 3–11. 2008.
57. *Biomimetics—Conception and strategy—Differences between biomimetic and conventional methods/products*, in *ISO/TC 266/SC, ISO/WD 18458*. 2013.
58. Gebeshuber, I., and Drack, M., An attempt to reveal synergies between biology and mechanical engineering. *In Proc. of the Institution of Mechanical Engineers*, Part C: *Journal of Mechanical Engineering Science*, **222**(7), p. 1281–1287. 2008.
59. Vattam, S.S., Helms, M.E., and Goel, A.K., Compound Analogical Design: Interaction between Problem Decomposition and Analogical Transfer in Biologically Inspired Design, *in Design Computing and Cognition.*, Springer. p. 377–396, 2008.
60. Van den Besselaar, P., and Heimeriks, G., Disciplinary, multidisciplinary, interdisciplinary: Concepts and indicators. *In International society of scientometrics and informetics conference (ISSI),* Sydney, Australia, 2001.
61. Full, R., TED: Learning from the gecko's tail [video], Retrieved from http://www.youtube.com/watch?v=d3syTrElgcg, 2009.
62. Reich, Y. and O. Shai, The interdisciplinary engineering knowledge genome. *Research in Engineering Design*, 23(3): p. 251–264, 2012.
63. Reich, Y., Shai, O., Subrahmanian, E., Hatchuel, A., Le Masson, P., The interplay between design and mathematics: Introduction to bootstrapping effects. *In Proc. of the 9th Biennial ASME Conference on Engineering Systems Design and Analysis* (ESDA), Haifa, 2008.
64. Parvan, M., Oepke, H., Kaiser, M.K., Lindemann, U., Role development for interdisciplinary collaboration support in biomimetics. *International Conference on Industrial Engineering and Engineering Management (IEEM),* Hong-Kong, 2012.
65. Schmidt, J.C., *Bionik und Interdisziplinarität. Wege zu einer bionischenZirkulationstheorie der Interdisziplinarität.*, in *Bionik. Aktuelle Forschungsergebnisse aus Nature, Ingenieur-und Geisteswissenschaften*, Berlin: Springer, p. 219–246, 2005.
66. http://www.swedishbiomimetics.com/. [cited 01.18.2014].
67. Reich, Y. and E. Subrahmanian. Designing PSI: An introduction to the PSI framework. in DS 80–2 Proceedings of the 20th International Conference on Engineering Design (ICED 15) Vol 2: Design Theory and Research Methodology Design Processes, Milan, Italy, 27-30.07. 15. 2015.

68. Subrahmanian, E., Reich, Y., Smulders, F., Meijer, S. Designing: insights from weaving theories of cognition and design theories. in DS 68–7: Proceedings of the 18th International Conference on Engineering Design (ICED 11), Impacting Society through Engineering Design, Vol. 7: Human Behaviour in Design, Lyngby/Copenhagen, Denmark, 15.-19.08. 2011.
69. Meijer, S., Reich, Y., and Subrahmanian, E. The future of gaming for design of complex systems. Back to the future of gaming. Gutersloh: W. Bertelsmann, 2014.
70. *VDI 2011 Biomimetics: conception and strategy—difference between bionic and conventional methods/products*, in *VDI 6220: 2011-06*, Berlin, 2011.
71. Sartori, J., Pal, U., and Chakrabarti, A., A methodology for supporting "transfer" in biomimetic design. *Artificial Intelligence for Engineering Design, Analysis and Manufacturing*, **24**(4), p. 483, 2010.
72. Mckeag, T., Framing your problem with the bio-design cube. *Zygote Quarterly*, (6), 2013.
73. http://dilab.cc.gatech.edu/dane/. [cited 02.10.2014].
74. Wilson, J.O., *A Systematic Approach to Bio-Inspired Conceptual Design*. PHD Thesis, Georgia Institute of Technology, 2008.
75. Vattam, S., Helms, M.E., and Goel, A.K., A content account of creative analogies in biologically inspired design. *Artificial Intelligence for Engineering Design, Analysis and Manufacturing*, **24**(4), p. 467–481, 2010.
76. Mak, T., and Shu, L., Using descriptions of biological phenomena for idea generation. *Research in Engineering Design*, **19**(1), p. 21–28, 2008.
77. Vosniadou, S., and Ortony, A., *Similarity and Analogical Reasoning*. Cambridge University Press, 1989.
78. Jacobs, S.R., Nichol, E.C., and Helms, M.E., Where Are We Now and Where Are We Going? The BioM Innovation Database. *Journal of Mechanical Design*,. 136(11): p. 111101, 2014.
79. Gramann, J., *Problemmodelle und Bionik als Methode*., PHD Thesis, Technical University Munich, 2004.
80. Alizadehbirjandi, E., Tavakoli-Dastjerdi, F., St Leger, J., Davis, S.H., Rothstein, J.P., Kavehpour, H.P. Ice Formation Delay on Penguin Feathers. *Bulletin of the American Physical Society*, 60, 2015.
81. http://amorphical.com/. [cited 10.01.2016].
82. Becher, T., and Trowler, P., *Academic Tribes and Territories: Intellectual* Enquiry and the Culture of Disciplines. McGraw-Hill International, 2001.
83. Shai, O. and Y. Reich, Infused design. I. Theory. Research in Engineering Design, 15(2): p. 93–107, 2004.
84. Shai, O. and Y. Reich, Infused design. II. Practice. Research in Engineering Design, 15(2): p. 108–121, 2004.
85. Nagel, R.L., Midha, P.A, Tinsley, A., Stone, R.B., McAdams, D.A., Shu, L.H., Exploring the use of functional models in biomimetic conceptual design. *Journal of Mechanical Design*, **130**, p. 121102, 2008.
86. Vattam, S., and Goel, A., Foraging for inspiration: Understanding and supporting the information seeking practices of biologically inspired designers. In *Proc. ASME Design Engineering Technical Conferences on Design Theory and Methods, Washington, DC,* 2011.
87. www.asknature.org. [cited 01.24.2014].
88. Haseyama, M., Ogawa, T., and Yagi, N., A Review of Video Retrieval Based on Image and Video Semantic Understanding. *ITE Transactions on Media Technology and Applications*, 1(1): p. 2–9, 2013.
89. Bruck, H.A., Gershon, A.L., Golden, I., Gupta, S.K., Gyger Jr L.S, Magrab, E.B., Spranklin, B.W., Training mechanical engineering students to utilize biological inspiration during product development. *Bioinspiration & Biomimetics*, **2**(4), p. S198. 2007.

90. Shu, L., A natural-language approach to biomimetic design. *AI EDAM (Artificial Intelligence for Engineering Design, Analysis and Manufacturing)*, **24**(4): p. 507, 2010.
91. Chiu, I., and Shu, L., *Natural language analysis for biomimetic design.* in *Proc. of ASME Design Engineering Technical Conferences and Computers and Information in Engineering Conference, Salt Lake City, Utah,* 2004.
92. Cheong, H., Chiu, I., Shu, L.H., Stone, R.B., McAdams, D.A., Biologically meaningful keywords for functional terms of the functional basis. *Journal of Mechanical Design*, **133**, p. 021007, 2011.
93. Stone, R.B., and Wood, K.L., Development of a functional basis for design. *Journal of Mechanical Design*, **122**, p. 359, 2000.
94. Pahl, G., and W. Beitz., *Engineering Design: a Systematic Approach.* Springer-Verlag, London, 1996.
95. Nagel, J.K., Stone, R.B., and McAdams, D.A., An engineering-to-biology thesaurus for engineering design. *in Proc. ASME IDETC/CIE, Montreal, Quebec, Canada*, 2010.
96. Nagel, J.K., and Stone, R.B., A computational approach to biologically inspired design. *Artificial Intelligence for Engineering Design, Analysis and Manufacturing*, **26**(02), p. 161–176, 2012.
97. Bohm, M.R., Vucovich, J.P., and Stone, R.B., Using a design repository to drive concept generation. *Journal of Computing and Information Science in Engineering*, **8**(1), p. 014502, 2008.
98. Lindemann, U., and Gramann J., Engineering design using biological principles. In *Proc. of the 8th International Design Conference,* 2004.
99. Hill, B., Goal setting through contradiction analysis in the bionics-oriented construction process. *Creativity and Innovation Management*, **14**(1), p. 59–65, 2005.
100. Yen, J., Weissburg, M., Helms, M., Goel, A., Biologically inspired design: a tool for interdisciplinary education, in *Biomimetics: Nature Based Innovation*, CRC Press/Taylor Francis, p. 332–356, 2011.
101. Bar-Cohen, Y., *Biomimetics: Reality, Challenges, and Outlook, in Biomimetics: biologically inspired technologies.* CRC Press, 2005.
102. Stone, R.B., Wood, K.L., and Crawford, R.H., A heuristic method for identifying modules for product architectures. *Design Studies*, **21**(1), p. 5–31, 2000.
103. Vakili, V., Chiu, I., Shu, L.H., McAdams, D. A., Stone, R., Including functional models of biological phenomena as design stimuli. *in Proc. International Design Engineering Technical Conferences & Computers and Information in Engineering Conference (ASME IDETC/CIE)*, 2007.
104. Goel, A.K., Vattam, S., Wiltgen, B., Helms, M., *Information-Processing Theories of Biologically Inspired Design, in Biologically Inspired Design*, Springer, p. 127–152, 2014.
105. Hubka, V., and Eder, W.E., *Theory of Technical Systems: a Total Concept Theory for Engineering Design.* Springer-Verlag, 1988.
106. Freitas Salgueiredo, C., Modeling inspiration for innovative NPD: lessons from biomimetics, *in 20th International Product Development Management Conference*, Paris, 2013.
107. Hatchuel, A., and Weil, B., A new approach of innovative design: an introduction to CK theory, *in Proc. of International Conference on Engineering Design*, 2003.
108. Vakili, V., and Shu, L., Towards biomimetic concept generation, *In Proc. of the ASME Design Engineering Technical Conference*, 2001.
109. http://www.mie.utoronto.ca/labs/bidlab/. [cited 01.24.2014].
110. Cheong, H., Chiu, I., and Shu L., Extraction and transfer of biological analogies for creative concept generation. in *Proc. ASME Int. Design Engineering Technical Conf. and Computers and Information in Engineering Conf*, 2010.
111. Goel, A.K., Rugaber, S., and Vattam, S., Structure, behavior, and function of complex systems: the structure, behavior, and function modeling language. *Artificial Intelligence for Engineering Design, Analysis and Manufacturing (AI EDAM)*, **23**(1), p. 23–35, 2009.

112. Chakrabarti, A., Sarkar, P., Leelavathamma, B., Nataraju, B.S., A functional representation for aiding biomimetic and artificial inspiration of new ideas. *Artificial Intelligence for Engineering Design, Analysis and Manufacturing (AI EDAM)*, **19**(2), p. 113–132, 2005.
113. Goel, A.K., Vattam, S., Wiltgen, B., Helms, M., Cognitive, collaborative, conceptual and creative—Four characteristics of the next generation of knowledge-based CAD systems: A study in biologically inspired design. *Computer-Aided Design*, **44**(10), p. 879–900, 2012.
114. Srinivasan, V., and Chakrabarti, A., *SAPPhIRE–an approach to analysis and synthesis. In International Conference on Engineering Design (ICED)*, Stanford, 2009.
115. Terninko, J., Zusman A., and Ziotin B., *Systematic Innovation: an Introduction to TRIZ (theory of inventive problem solving)*, CRC press, 1998.
116. Altsuller, G.S., Shulyak, L., Rodman, S., Fedoseev, U., *40 Principles: TRIZ keys to technical innovation*, Technical Innovation Center, Inc, 1997.
117. Bertalanffy, L.V., *General System Theory. George Braziller*, Inc. 1969.
118. Zavbi, R., and Duhovnik, J., Prescriptive model with explicit use of physical laws, in *International Conference on Engineering Design*, Tampere, 1997.
119. Vincent, J.F., and Mann, D.L., Systematic technology transfer from biology to engineering. *Philosophical Transactions of the Royal Society of London. Series A: Mathematical, Physical and Engineering Science*, **360**(1791), p. 159–173, 2002.
120. Cross, N., *Developments in design methodology*. John Wiley & Sons, 1984.
121. Reich, Y., Layered models of research methodologies. *Artificial Intelligence for Engineering, Design, Analysis and Manufacturing*, 8(4), p. 263–274, 1994.
122. Kolberg, E., Reich, Y., and Levin, I., Designing winning robots by careful design of their development process. *Research in Engineering Design*, **25**(2), p. 157 183, 2014.
123. Reich, Y., My method is better!. *Research in engineering design*, **21**(3), p. 137–142, 2010.
124. Shai, O., Reich, Y., Hatchuel, A., Subrahmanian, E., Creativity and scientific discovery with Infused Design and its analysis with C-K theory, *Research in Engineering Design*, **24**(2), 201–214, 2013.
125. Nunamaker, J.F., Chen, M., and Purdin, T.D.M., Systems development in Information System Research, *Journal of Management Information Systems Research*, **7**(3), p. 89–106, 1991.
126. Reich, Y., Subrahmanian, E., Cunningham, D., Dutoit, A., Konda, S., Patrick, R., Westerberg, A., Building agility for developing agile design information systems. *Research in Engineering Design*, **11**(2), p. 67–83, 1999.
127. Eisenhardt, K.M., Building theories from case study research. *The Academy of Management Review*, **14**(4), p. 532–550, 1989.
128. Yin, R.K., *Case Study research: Design and Methods*, Newbury Park: Sage Publications, 2003.
129. www.biomimicrynews.org. [cited 01.18.2014].
130. Kauffman, D., *Introduction to Systems Thinking*, Future Systems/Tlh Assoc, 1981.
131. Meadows, D.H., *Thinking in Systems, A primer*. Chelsea Green Publishing, 2008.
132. Altshuller, G.S., *Creativity as an Exact Science: The Theory of the Solution of Inventive Problems*, Vol. 5, CRC Press, 1984.
133. Rosse, C., and Mejino, J., A reference ontology for bioinformatics: The Foundational Model of Anatomy. *Journal of Biomedical Informatics*, **36**, p. 478–500, 2003.
134. Ariew, A., Cummins, R., and Perlman, M., *Functions: New Essays in the Philosophy of Psychology and Biology*. New York: Oxford University Press, 2002.
135. Stone, R. B., Wood, K.L., Development of a functional basis for design. *Journal of Mechanical Design*, **122**(4), p. 359–370, 2000.
136. Cascini, G., Rotini, .F., Russo, D., Functional modeling for TRIZ-based evolutionary analysis. In *International Conference on Engineering Design* (ICED), CA, USA, 2009.
137. Hirtz, J., Stone, R.B., McAdams, D.A., Szykman, S., Wood, K.L., A functional basis for engineering design: reconciling and evolving previous efforts. *NIST Technical Note 1447*, 2002.

138. Hirtz, J., Stone, R.B., McAdams, D.A., Szykman, S., Wood, K.L., A functional basis for engineering design: reconciling and evolving previous efforts. *Research in engineering Design*, **13**(2), p. 65–82, 2002.
139. Burek, P., *Ontology of Functions: A Domain Independent Framework*, PHD Thesis, University of Leiptzig, 2006.
140. Russo, D., Regazzoni, D., and Montecchi, T., Eco-design with TRIZ laws of evolution. *Procedia Engineering*, **9**, p. 311–322, 2011.
141. D'Anna, W., and Cascini, G., Supporting sustainable innovation through TRIZ system thinking. *Procedia Engineering*, **9**, p. 145–156, 2011.
142. Bukhman, I., *TRIZ Technology for Innovation*, Cubic Creativity Company, 2012.
143. Salamatov, Y., *TRIZ: the Right Solution at the Right Time: a guide to innovative problem solving*. The Netherlands: Insytec BV, Hattem, 1999.
144. Mann, D., *The TRIZ route to naturally better system design.* Systematic innovation.
145. Terninko, J., Zusman, A.,. Zlotin, B., *Systematic Innovation: An Introduction to Triz*. St. Lucie Press, 1998.
146. Buckl, S., Matthes, F., Schweda, C. M., Utilizing patterns in developing design theories, in *International Conference on Information Systems,* 2010.
147. Alexander, C., *Notes of the Synthesis of Form*. Vol. 5, Harvard University Press, 1964.
148. Alexander, C., Ishikawa, S., and Silverstein, M., *A Pattern Language: Towns, Buildings, Construction*. London: Oxford University Press, 1977.
149. Gamma, E., Helm, R., Johnson, R. E., Vlissides, J. M., *Design Patterns: Elements of Reusable Object-Oriented Software*, Munich, Germany: Addison-Wesley Professional, 1994.
150. DeMarco, T., Hruschka, P., Lister, T., Robertson, S., Robertson, J., McMenamin,S, Junkies, A., and Zombies, T., *Understanding Patterns of Project Behavior*. New York, NY, USA: Dorset House, 2008.
151. Buschmann, F., Meunier, R., Rohnert, H., Sommerlad, P., Stal, M., *Pattern Oriented Software Architecture: a System of Patterns*. New York, NY, USA: John Wiley & Sons, 1996.
152. Horowitz, R., *Creative Problem Solving in Engineering Design*. PHD Thesis, Tel Aviv University, 1999.
153. Suh, N.P., *The Principles of Design*. Oxford University Press, 1990.
154. Goel, A.K., and Bhatta, S.R., Use of design patterns in analogy-based design. *Advanced Engineering Informatics*, **18**(2), p. 85–94, 2004.
155. Bhatta, S.R., and Goel, A.K., A functional theory of design patterns. *In International Joint Conference on Artificial Intelligence (IJCAI), (1)*, Nagoya, Japan, 1997.
156. Ball, P., *Shapes-Nature's Patterns a Tapestry in Three Parts*. Oxford University Press, 2009.
157. Foy, S., *The Grand Design: Form and Colour in Animals*. Prentice-Hall, Inc, 1983.
158. Haeckel, E., *Art Forms in Nature: The Prints of Ernst Haeckel: One Hundred Color Plates*. Prestel Pub, 1998.
159. Pearce, P., *Structure in Nature is a Strategy for Design*. The MIT Press, 1978.
160. Stevens, P.S., *Patterns in Nature*. Atlantic Monthly Press, 1974.
161. Thompson, D., *On Growth and Form*. Cambridge University Press, 1966.
162. Tsui, E., *Evolutionary Architecture-Nature as a Basis for Design*. New York: John Wiley & sons, 1999.
163. Feynman R. P., Leighton, R.B., and Sands, M., *The Feynman Lectures on Physics*. Vol. 2., Massachusetts: Addison-Wesley Publishing Company, Inc, 1964.
164. Vogel, S., *Life's Devices-The Physical World of Animals and Plants*. Princeton University Press, 1988.
165. Turner, J.S., *The Tinkerer's Accomplice: How Design Emerges from Life Itself.* Harvard University Press, 2009.

166. Lauder, G.V., Functional morphology and systematics: studying functional patterns in historical context. *Annual Review of Ecology and Systematics*, **21**, p. 317–340, 1990.
167. Koehl, M.A.R., When does morphology matter?. *Annual Review of Ecology and Systematics*, **27**, p. 501–542, 1996.
168. Ritchey, T., Modelling complex socio-technical systems using morphological analysis. *Swedish Morphological Society,* Stockholm, 2002.
169. Vincent, J., Biomimetics—a review. Proceedings of the Institution of Mechanical Engineers, Part H: *Journal of Engineering in Medicine*,. 223(8): p. 919–939, 2009.
170. Gorb, S.N., Functional Surfaces in Biology: Mechanisms and Applications, in *Biomimetics: Biologically Inspired Technologies*, CRC Press: Boca Raton, FL, p. 381–397. 2006.
171. Gorb, S., *Attachment Devices of Insect Cuticle*. Springer, 2001.
172. Helfman, C.Y., Reich, Y., Greenberg. S., Substance field analysis and biological functions, in *ETRIA TRIZ future conference,* Lisbon, 2012.
173. Altshuller, G.S., Zlotin.B.L., Zusman, A.V., Filatov, V.I., *Search for New Ideas: from Insight to Technology (Theory and Practice of Inventive Problem Solving)*, Kishinev: Kartia Mldoveniaske, 1989.
174. Zwicky, F., *The morphological approach to discovery, invention, research and construction*, in *New Methods of Thought and Procedure*. Springer, p. 273–297, 1967.
175. Hundal, M., A systematic method for developing function structures, solutions and concept variants. *Mechanism and Machine theory*, **25**(3), p. 243–256, 1990.
176. Helfman, C.Y., Reich, Y., Greenberg. S., What can we learn from biological systems when applying the law of system completeness?. i*n ETRIA TRIZ future conference*, Dublin, 2011.
177. Autumn, K., Liang, Y.A., Hsieh, S.T., Zesch, W., Chan, W.P., Kenny, T.W., Fearing, R., Full, R., Adhesive force of a single gecko foot-hair. *Nature*, 405 (6787), p. 681–685, 2000.
178. Evans, M., The jump of the click beetle (Coleoptera, Elateridae)—a preliminary study. *Journal of Zoology*, **167**(3), p. 319–336, 1972.
179. Smith, B.L., Schäffer, T.E., Viani, M., Thompson, J.B., Frederick, N.A., Kindt, J., Belcher, A., Stucky, G.D., Morse, D.E., Hansma, P.K., Molecular mechanistic origin of the toughness of natural adhesives, fibers and composites. *Nature*, 399(6738), p. 761–763, 1999.
180. Helfman, C.Y., Reich, Y., Greenberg. S., Biomimetics: Structure-function patterns approach. *Journal of Mechanical Design*, 136(11), 111108–1, 2014.
181. Beisel, D.E.A., *Bionics-Nature Patents*. Pro Futura, 1991.
182. Tributsch, H., *How Life Learned to Live: Adaptation in Nature*. Mit Press, 1982.
183. Seki, Y., Schneider, M.S., and Meyers, M.A., Structure and mechanical behavior of a toucan beak. *Acta Materialia,*. **53**(20): p. 5281–5296, 2005.
184. Fratzl, P. and Weinkamer, R., Nature's hierarchical materials. *Progress in Materials Science*, **52**(8), p. 1263–1334, 2007.
185. Scarr, G., Helical tensegrity as a structural mechanism in human anatomy. *International Journal of Osteopathic Medicine,* **14**(1), p. 24–32, 2011.
186. Wainwright, S., Vosburgh, F., and Hebrank, J., Shark skin: function in locomotion. *Science*, **202**(4369), p. 747–749, 1978.
187. Kirschner, C.M., and Brennan, A.B., Bio-inspired antifouling strategies. *Annual Review of Materials Research*, **42**, p. 211–229, 2012.
188. Lev-Yadun, S., Katzir, G., and Neeman, G., Rheum palaestinum (desert rhubarb), a self-irrigating desert plant. *Naturwissenschaften*, **96**(3), p. 393–397, 2009.
189. Quick, D., *Filter feeding basking shark inspires more efficient hydroelectric turbine*. 2011. available in webpage http://www.gizmag.com/strait-power-hydroelectric-turbine/17801/, [accessed 24 January, 2013].
190. Dawson, C., Vincent, J.F., and Rocca A.M., How pine cones open. *Nature*, **390**(6661), p. 668–668, 1997.
191. Wainwright, P.C., Turingan, R.G., and Brainerd, E.L., Functional morphology of pufferfish inflation: mechanism of the buccal pump. *Copeia*, p. 614–625, 1995.

192. Cassidy, J., Hiltner, A., and Baer, E., Hierarchical structure of the intervertebral disc. *Connective Tissue Research,* **23**(1), p. 75–88, 1989.
193. www.findstructure.org. [accessed 24 January, 2016].
194. Vogel, S., *Cats' Paws and Catapults: Mechanical Worlds of Nature and People*. WW Norton & Company, 2000.
195. Coppola, G., and Caro, C., Oxygen mass transfer in a model three-dimensional artery. *Journal of The Royal Society Interface*, **5**(26), p. 1067–1075, 2008.
196. Arzt, E., Gorb, S., and Spolenak, R., From micro to nano contacts in biological attachment devices. *In Proc. of the National Academy of Sciences*, **100**(19), p. 10603–10606, 2003.
197. Helfman, C.Y., Reich, Y., Greenberg. S., Sustainability strategies in nature, in *7th Design & nature* conference, Opatja, 2014.
198. Chen, J.L., and Yang, Y-C., Eco-Innovation by Integrating Biomimetic with TRIZ Ideality and Evolution Rules, in *Glocalized Solutions for Sustainability in Manufacturing*, Springer. p. 101–106, 2011.
199. TRIZ Solutions LLC, Law of technical system evolution Presentation.
200. Oksman-Caldentey, K.M., and Barz, W.H., *Plant Biotechnology and Transgenic Plants*. Vol. 92, CRC press, 2002.
201. Whittlesey, R.W., Liska, S., and Dabiri, J.O., Fish schooling as a basis for vertical axis wind turbine farm design. *Bioinspiration & Biomimetics*, **5**(3), p. 035005, 2010.
202. Weiss, A., Iko, A., Helfman Cohen, Y., Das Amarendra, K., Mazor, G., The ideality what model for product design, in 17th International conference on engineering and product design education, Loughborough, 2015.
203. Ashley, S., Designing for the environment. Mechanical Engineering,. 115(3): p. 52, 1993.
204. Robotham, A.J., The use of function/means trees for modelling technical, semantic and business functions. *Journal of Engineering Design*, **13**(3), p. 243–251, 2002.
205. Andreasen, M.M., *Machine Design Methods Based on a Systematic Approach-Contribution to a Design Theory.* PHD Thesis, Department of Machine Design, Lund University, Sweden, 1980.
206. Belski, I., *Improve your thinking: substance-field analysis*, TRIZ4U, 2007.
207. Hyde, E., The function of the hilum in some Papilionaceae in relation to the ripening of the seed and the permeability of the testa. *Annals of Botany*, **18**(2), p. 241–256, 1954.
208. Sanggaard, K. W., Danielsen, C.C., Wogensen, L., Vinding, M.S., Rydtoft, L.M., Mortensen, M.B., Karring, H., Nielsen, N.C., Wang, T., Thøgersen, I.B., Unique structural features facilitate lizard tail autotomy. *PloS one*, **7**(12), p. e51803, 2012.
209. Gupta, B., Investigations of the rolling mechanism in the Indian hedgehog. *Journal of Mammalogy*, p. 365–371, 1961.
210. www.calcalist.co.il/Ext/Comp/ArticleLayout/CdaArticlePrintPreview/1,2506,L-3583146,00.html. [cited 01.25.14].
211. Moon, B.R., The mechanics of swallowing and the muscular control of diverse behaviours in gopher snakes. *Journal of Experimental Biology*, **203**(17), p. 2589–2601, 2000.
212. Nelson, B., Wilson, J., and Yen J., A study of biologically-inspired design as a context for enhancing student innovation. in *Frontiers in Education Conference, 2009.*
213. Shah, J.J., Smith, S.M., and Vargas-Hernandez, N., Metrics for measuring ideation effectiveness. *Design Studies*, **24**(2), p. 111–134, 2003.
214. Domino, G., Assessment of cinematographic creativity. *Journal of Personality and Social Psychology*, **30**(1), p. 150, 1974.
215. Moulianitis, V., Aspragathos, N., and Dentsoras, A., A model for concept evaluation in design–an application to mechatronics design of robot grippers. *Mechatronics*, **14**(6), p. 599–622, 2004.
216. Cropley, D., and Cropley, A., *Engineering Creativity: A Systems Concept of Functional Creativity*. Lawrence Erlbaum Associates Publishers, 2005.
217. Amabile, M., Social Psychology of Creativity: A Consensual Assessment Technique. *Journal of Personality and Social Psychology*, **43**(5), p. 997–1013, 1982.

218. Bruner, J.S., The Conditions of Creativity. In H. Gruber, G. Terrell, & M. Wertheimer (Eds.), *Contemporary Approaches to Creative Thinking*. New York: Atherton Press, 1962.
219. Oman, S.K., Tumer, I. Y., Wood, K., Seepersad, C., A comparison of creativity and innovation metrics and sample validation through in-class design projects. *Research in Engineering Design*, **24**(1), p. 65–92, 2013.
220. Howard, T., Culley, S., and Dekoninck, E., Creative stimulation in conceptual design: an analysis of industrial case studies. in *ASME International Design Engineering Technical Conferences & Computers and Information in Engineering Conference (IDETC/CIE)*, University of Bath, 2008.
221. Christensen, B.T., and Schunn, C.D., The relationship of analogical distance to analogical function and preinventive structure: The case of engineering design. *Memory & Cognition*, **35**(1), p. 29–38, 2007.
222. Schmidt-Nielsen,K., Taylor, C., and Shkolnik A., Desert snails: problems of heat, water and food. *Journal of Experimental Biology*, **55**(2), p. 385–398, 1971.
223. Barthlott, W., Schimmel, T., Wiersch, S., Koch, K., Brede, M., Barczewski, M., Walheim, S., Weis, A., Kaltenmaier, A., Leder, A., The Salvinia paradox: superhydrophobic surfaces with hydrophilic pins for air retention under water. *Advanced Materials*, **22**(21), p. 2325–2328, 2010.
224. Schimmel, T. *The Salvinia Effect.*http://aph-ags.webarchiv.kit.edu/salvinia-e.html. [cited 01.25.2014].
225. Hatchuel, A., and Weil, B., C-K design theory: an advanced formulation. *Research in engineering design*, **19**(4), p.181–192, 2009.
226. Reich, Y., Hatchuel, A., Shai, O., and Subrahmanian, E., A theoretical analysis of creativity methods in engineering design: casting and improving ASIT within C–K theory. *Journal of Engineering Design*, **23**(2), p.137–158, 2012.

Index

Y.H. Cohen and Y. Reich, *Biomimetic Design Method for Innovation and Sustainability*, DOI 10.1007/978-3-319-33997-9

Printed in the United States
By Bookmasters